Biofuels Manufacture from Renewables other than Corn Starch

by Matthew Katterman

ISBN-13 : 978-1514143518

ISBN-10: 1514143513

Copyright © 2015 Matthew Katterman

1st Edition

This book is dedicated to my mother Grace Lenore Katterman for the invaluable support she has given to me while writing this manuscript. Thank you for your help.

Contents:

Introduction	5
1 : Disadvantages of Corn Starch Ethanol Production	11
2 : The Emerging Biofuel Infrastructure	37
3 : Renewable Fuels from Thermochemical Processing	69
4 : Biodiesel Manufacture	95
5 : Alternative Ethanol Production Methods	131
6 : Mixed Alcohol Production	157
7 : Biofuels from Algae	185
Glossary	217

Introduction

Currently, two types of biofuels are implemented on a large scale volume production in the United States. These biofuels are ethanol and biodiesel. Biodiesel within the United States is being manufactured at around 1 billion gallons per year. Ethanol is produced at much higher volumes in the US as it is estimated that 15 billion gallons of ethanol will be needed each year for use as a gasoline fuel additive (E10). Ethanol is a fuel additive primarily used to boost octane performance in a spark ignition engine.

Ethanol is made during a fermentation process done by yeasts or other microbes that essentially convert saccharides (i.e., sugars) into alcohol (ethanol). Alcoholic fermentation has its roots from the manufacture of beverages such as beer and wine produced many centuries ago originating from the Far East and Middle Eastern areas of the world. It has been known that microbe cells can ferment a wide variety of plant sources such as grapes, dates, molasses, barley, etc into ethanol.

In our recent American History, poor farmers would make extra revenue by producing ethanol in stills hidden in the back woods and then deliver the product to customers in modified cars. People known as moonshiners or bootleggers cultivated adequate quantities of ethanol and evaded government officials as revenuers sought it important to collect proceeds from ethanol sales. Ethanol in the past has also been used as a vehicle fuel in early internal combustion engines as it was one of the primary vehicle fuel sources utilized in these engines dating back to the early 1800's.

In our very recent US history, ethanol was chosen as a replacement fuel additive to MTBE. Due to the environmental concerns of soil, water and air contamination, the widespread mix of MTBE into gasoline was prohibited in many states by 2004. Ethanol made from corn starch was the more logical feedstock source due to existing technology and the easier ability to ferment starch sources versus alternative carbohydrates sources like cellulose. Most, if not all of this amount of ethanol is manufactured from corn starch refineries.

To date in the United States, corn starch ethanol manufacture is the main method of ethanol production that is being done at a large volume capacity and at an affordable cost. However, this would not have been possible if not for the government support in the form of various tax breaks, credits and loans that assisted the manufacture of corn starch ethanol within a short period of time (i.e., circa 2004 and later).

Corn's rapid, widespread cultivation has also brought up concerns towards its use as a biofuel when it is also used as a food crop. Corn's increased cultivation, in part, has caused other food commodities to increase in price as well. An increase in corn cultivation also has the tendency to compete with other food crops for farmland space within the United States. Fifteen billion gallons of ethanol manufactured per year with corn would use a significant amount of farmland in the US. An array of issues and policies regarding the production of ethanol from corn starch are discussed in the introductory chapter (chapter 1) of this book.

Although the majority of people in the US are unaware of it, ethanol's practical use in our vehicles and in the current fuel infrastructure itself still faces many obstacles that may be difficult to overcome. These issues are mainly related to its inherent chemical properties that are common amongst low molecular weight oxygenated organic compounds.

However, it is felt that the production of ethanol itself should not necessarily be discontinued due to these shortcomings. For example, it may be shown in the future that hydrous ethanol would be more advantageous to utilize as a fuel additive in gasoline and may be easier to transport across the country when compared to the anhydrous version we currently produce. Ethanol as a chemical compound is also suited for other potential energy applications and further chemical production. For example, ethylene, the most important petrochemical building block utilized for plastics, can be derived from ethanol. Ethanol is also a main ingredient required to make other biofuels such as biodiesel or ether compounds.

One of the goals of this publication is to discourage the continued practice of producing ethanol or other fuels from corn starch and instead turn towards alternate biofuel manufacturing technologies that implement other feedstock sources such as algae and lignocellulosics. It is felt that enough resources obtained from lignocellulosic wastes, energy crops and algae cultivation exist in order to meet vehicle fuel demands not only in the United States but for the world as a whole as well.

Therefore, the material in this publication in large part concentrates on the production of ethanol, other alcohols, biodiesel, renewable diesel, gasoline and jet fuels from the large supplies of renewable feedstock sources that we have in the US. These sources can be obtained from our normal agricultural or municipal waste streams generated on a regular basis. Recycling urban based municipal wastes would also contribute towards environmental remediation efforts by helping to reduce unnecessary landfill space. In addition, the growth of energy crops on marginal or non-competing lands along with cultivation of algae could greatly boost biofuel's overall production capacity. Therefore, it is felt that utilizing these types of renewable feedstock sources may be more conducive towards manufacturing biofuels at a local level since feedstock supply is not necessarily limited to a regional location like corn is. This practice would also lessen the need to transport the biofuel great distances across the country.

Alternative biofuel production usually falls into one of two categories, straight hydrocarbon manufacture or production of oxygenated organic compounds such as alcohols. This book attempts to cover the manufacturing technologies concerning both types. Oxygenated biofuel compounds for the most part in modern times are starting to apply as fuel additives to either gasoline or diesel fuel. However, if these compounds have the correct composition and properties they could apply towards use as a stand alone fuel supply as well. For example, biodiesel can serve as a fuel additive or stand alone fuel. However, even as a stand alone fuel biodiesel may require other fuel additives in order to be utilized effectively as a diesel fuel.

Although biodiesel experienced a recent short lived period of widespread amateur based production, it is felt that this fuel still has its place in the renewable diesel fuel based infrastructure and market. Several issues regarding biodiesel manufacture are addressed in this book such as sources and supply of biodiesel, effectiveness as a fuel additive or stand alone fuel, its beneficial and adverse chemical based properties and consideration of a waste glycerine market. Biodiesel is covered in more detail in chapter 4 of this book.

There is also an increasing need towards manufacturing oxygenated biofuel compounds other than just ethanol or biodiesel. It is known that a certain percentage of our hydrocarbon based fuel supply should contain some oxygen in it for vehicle performance and engine efficiency reasons. Any type of legitimately established oxygenated compound can fulfill this requirement. Therefore, the reader should understand that there is no reason why we should be locked into using ethanol as our sole fuel additive or even as our main source of biofuel in the future. In addition, it's important to understand that these other oxygenated fuel compounds can be made from renewable resources instead of relying upon petroleum or corn production.

In this book special attention is given to thermochemically based technologies that generate straight hydrocarbons utilized for biofuel. These technologies include thermochemical gasification, pyrolysis and liquefaction. More of this material is covered in chapter 3 of this book. It is also important to keep in mind that algae could serve as a major feedstock source applied towards making hydrocarbons from thermochemical processing. In fact throughout this book the flexibility of utilizing algae as a feedstock source has been demonstrated.

As far as straight hydrocarbon biofuel production is concerned, renewable diesel happens to be the most prevalent type of biofuel that can be produced from biomass sources. The production of renewable diesel takes place with all types of thermochemical processing methods. In addition, the emission of organic compounds from algae or microbes tends to be more suited towards renewable diesel production. Therefore, someday we may witness a revamped biofuel infrastructure that is based upon renewable diesel production.

Biofuels are principally produced at large volume levels in refineries. The goal of a modern biorefinery should be the manufacture of a biofuel done at a reasonable cost (around $1 or less per gallon), at a reasonably high production capacity (i.e., millions of gallons per year) while doing this energy efficiently and conserving on resources (i.e., water use). Examples of where modern biorefineries may already excel in these manufacturing based areas are the production of alternative ethanol from lignocellulosics. In chapter 5 of the book the reader will encounter some examples of alternative ethanol biorefineries that actually fulfill the aforementioned capabilities (energy efficiency, reasonable cost, high capacity, etc). These biorefineries are associated with the US DOE Integrated Biorefinery program. They are superior in energy efficiency, biofuel yield and water usage when compared to straight fermentative alcohol production. They are also known for their greater versatility in utilizing feedstock sources applied towards biofuel manufacture.

Therefore, it's important to understand that our US government has been very much involved with integrated biorefineries by allowing a variety of companies to come forth and prove that alternative biofuel production is possible at the pilot or demonstration plant level through government based grants or loans. As previously mentioned, this book covers some of the 2009 Integrated Biorefinery Demonstration program grant award recipients. It also needs to be understood that these emerging companies will most likely require more support and capital from government and private resources so that they can be successful in manufacturing alternative biofuels at high volume levels.

These newer biorefineries should also be able to help solve the carbon dioxide mitigation problem as it pertains to the reduction or capture of carbon dioxide emissions. In fact integrated biorefineries are required by mandate to reduce carbon dioxide emissions compared to the larger amounts emitted by petroleum based vehicle fuel refineries. They are able to do this through advanced carbon capture technologies such as pressure swing adsorption (PSA) or the use of industrial capturing units like MDEA or Selexol.

The buildup and construction of integrated biorefineries would also greatly enhance a biofuel based infrastructure. Similar to petroleum based vehicle fuel manufacture, biofuel manufacture has a very large potential to affect industries in other manufacturing areas through the production of biochemicals that can be utilized as industrial chemical building blocks. In addition, different from petroleum based manufacture, biofuel manufacture has the potential to beneficially contribute towards further food and agricultural production while also helping such industries focus on environmental remediation and recycling.

The big buzz word in modern biofuel manufacture is biocrude and like its name denotes, it happens to be the raw material similar in function to petroleum crude. Therefore, biocrude similar to petroleum crude can be converted into fuels as well as chemicals or other valuable byproducts. Biocrude manufacture has great potential when associated with algae cultivation. It allows for a wider conversion of the biomass into biofuel since several of the biocomponents in it can be converted into other organic compounds fashionable for further fuel use.

The last chapter discusses renewable biofuels made from algae. These biofuels include alcohols and renewable diesel. It focuses on the increased capability and versatility of algae as a feedstock source to produce biofuel since the biocomponents of lipids, saccharides and proteins can be converted into biofuel compounds through a variety of processing methods. One major factor affecting this type of production has to do with the manner in which the algae are originally cultivated before the actual biofuel processing takes place. In fact one could assert that the type of algae cultivation chosen principally determines what methods of processing and refining are necessary to turn it into biofuel.

Algae as a feedstock source also has great potential in producing oxygenated compounds such as alcohols. Alcohols can either be produced directly from the algae itself or are made from the fermentation of the sugars contained in the algae by microbes. Algae directly producing alcohols are an upcoming technology that should take place with many more production methods being available in the future.

Chapter 1

The Disadvantages of Corn Starch Ethanol Production

1.1 Our main supply of renewable biofuel comes from corn starch while cellulosic ethanol is also expected to contribute a large amount of biofuel in the future

Ethanol as a biofuel has been implemented as an additive to gasoline during the last several years, usually mixed at 10 % by volume. This gasoline based fuel mixture is known as E10. Since 2005, large amounts of ethanol are being manufactured from biorefineries that mainly utilize corn as the feedstock source. As Table 1-1 demonstrates, at least 13 billion gallons of ethanol has been produced every year in the US since 2010.[1] In order to fulfill E10 demands across the country the production of 15 billion gallons of ethanol are required. During the next decade the production of corn starch based ethanol is expected to increase up to 15 billion gallons of ethanol per year and be maintained at this capacity for years afterward.[2]

US ethanol production per year from corn-starch fermentation

Year	Million Gallons of Ethanol
2005	3904
2006	4855
2007	6500
2008	9000
2009	10600
2010	13298
2011	13948
2012	13300
2013	13312
2014	14340

Table 1-1 : Ethanol Production per year from 2005 from corn resources[1]

However, the current practice of producing ethanol from corn raises concerns since it primarily serves as a food crop. One concerning issue is whether too much farmland is being allocated for corn cultivation just to satisfy this ethanol demand. If this practice continues it becomes questionable as to whether our farmland resources would be able to support such a high volume of ethanol obtained from corn starch. For this reason it would be preferable to implement alternative feedstock sources towards producing ethanol.

The manufacture of ethanol from sources other than corn starch would help to satisfy this demand for ethanol. The government has already anticipated that future ethanol production be made from cellulosic sources like switchgrass. However, it expects that in the future both corn starch and cellulosic based ethanol production take place concurrently. According to the US Renewable Fuel Standards (RFS2), 16 billion gallons of ethanol will be made from cellulosic sources by 2022.[2] Future projections estimate the gradual ramping up of cellulosic ethanol production to the amount of 4 billion gallons per year increase every 3 years starting from 2016 as shown in Table 1-2.[2] In addition, corn starch ethanol production will be maintained at a production capacity of 15 billion gallons per year from 2016 to 2022.[2]

However, the question now becomes: why should we keep producing ethanol from corn if other feedstock sources such as lignocellulosics (cellulosic ethanol) can take its place? Recent government studies demonstrate this point. For example, a specific DOE study estimated that it may be possible to produce 60 billion gallons of ethanol from sources like switchgrass by the year 2030.[3]

Expected corn-starch & cellulosic ethanol production up through 2022

Year	Cellulosic Ethanol	Corn Starch Ethanol * Renewable Biofuel
2016	4 Billion Gallons	15 Billion Gallons
2019	8 Billion Gallons	15 Billion Gallons
2022	16 Billion Gallons	15 Billion Gallons

Table 1-2 : Renewable Fuel Standards as per the RFA estimated in the next decade[2]
* Renewable Biofuel mainly defined as ethanol derived from corn starch

1.2 Main food commodity crops that compete for land resources have the effect of raising the commodity price values

The major food commodity crops of corn, soybeans and wheat make up around two thirds of the total US farmland acreage. However, when they compete for farmland space, commodity crop prices tend to rise as a result. In the past, corn and soybeans took up about the same amount of farmland acreage, but recently that situation has changed. During the spring of 2007 the

USDA reported that the acreage of crops like soybeans decreased markedly while corn acreage soared to an all time high since 1944 due to an increased need for ethanol.[4] Two periods of major food commodity price rises have subsequently taken place during the years of 2005-2007 and 2009-2012.

Food prices usually go through cyclical periods of rise and falls. However, these changes have been dramatic during the last decade. The three major food commodity crops of corn, wheat and soybeans prices have doubled or even tripled in value before trying to return to previous price levels. This trend is shown in figure 1.1 for these commodity crops. The figure shows a history of price per bushel between the major food crops from 2005 – 2014.[5] There were two major price spike periods going from 2005 – 2007 and then again in 2009 – 2013, as was previously stated above. Only within the last few years (2013 – 2014) have prices began to return to normal values. What is intesting about these price changes is that they seem to affect all of the crops, not just one of them. Therefore, one could assert that they may be more sensitive to each other when changes in resource availability take place.

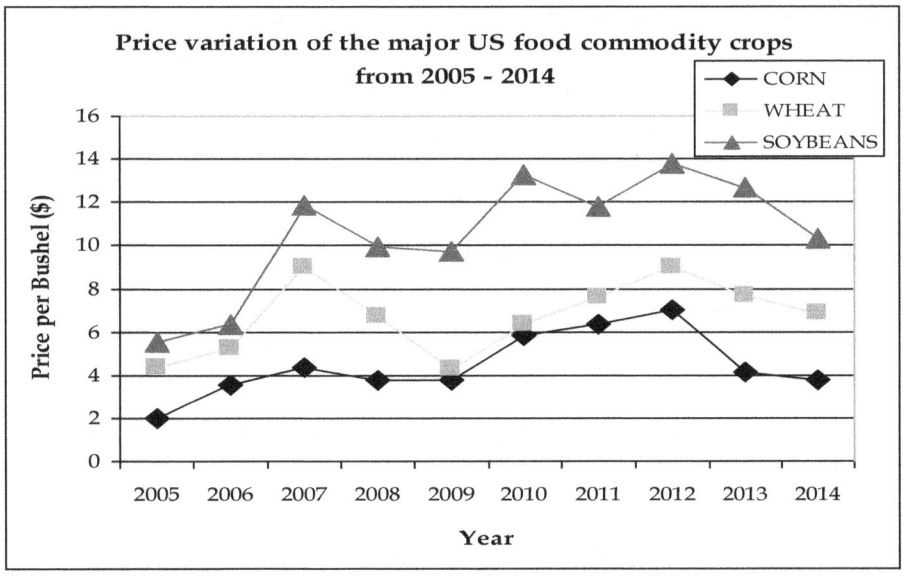

Figure 1-1 : Price per bushel of corn, wheat and soybeans during the years of 2005 – 2014 as per USDA statistics.[5]

Our last most recent food price spike that occurred during 2010 to 2012 alarmed the international community so much that it elicited a response from at least 10 world wide organizations claiming that our biofuel production caused overall world food prices to rise as well as bring about world food supply shortages.[6] For example, in the US alone it has been estimated that higher than normal corn production caused a 10 – 30 % price increase of food related commodities such as chicken, pork, eggs, breads, cereals and milk[7], which also require farmland related resources.

In addition, it is imperative that land set aside for corn cultivation not compete or take away from farmland set aside for growing wheat. For food sustainability purposes wheat currently happens to be the third largest food crop cultivated in the US. It required significantly more farmland for many years prior to the turn of the century (i.e., year 2000). Therefore, it's important that wheat maintain significant crop production yields obtained from its cultivation on adequate amounts of land. The amount of allocated farmland and yield per acre for the three major food commodity crops cultivated during 2011 is shown in Table 1-3.[8,9] Wheat utilizes less farmland acreage than corn or soybeans.

According to Table 1-3 it should also be noted that corn produces a very high biomass yield (~150 bushels per acre) while attaining the largest amount of farmland acreage of any food crop in the United States. The theoretical situation of relying upon corn exclusively for our renewable biofuel needs brings up concerns associated with the total amount of land required for such an endeavor. As an example, the amount of land acreage necessary to make ethanol the biofuel source of the world would take up least 10 % of the total terrestrial earth land area.[10]

US farmland acreage & yield per acre for the top three food crops in 2011

Crop Type	Total Acreage Amount	Yield per Acre
Corn	84 million acres	147 bushels / acre
Soybeans	74 million acres	42 bushels / acre
Wheat	46 million acres	44 bushels / acre

Table 1-3 : Amount of farmland acreage for the top three cultivated crops in the United States as of 2010.[8,9]

1.3 The main value component of food-energy crops should be less utilized for biofuel purposes

Food-energy crops can be defined as land crops used for the dual purpose of producing food commodities as well as manufacturing energy in the form of vehicle fuels or electrical energy. Food-energy crops can be useful in solving our energy problems when the leftover crop biomass called residuals is used for energy production. However, when the main value component of the crop competes for purpose as a food versus a biofuel commodity, it can lead to an increase in price for that crop.

The main value component of starch from corn is applied towards many consumable food or industrial products as well as ethanol for biofuel. Corn starch related products of value can be foods, cereals, beverages, sugar type products and industrial based products. However, unfavorable competition between consumable products and biofuel result from using the main value component of starch from corn. This can have the effect of raising food related consumable commodities such as corn based raw sugar products. The onset of high corn prices that took place during the years of 2004-2007 seems likely to have caused the prices of corn sugar products such as high fructose corn syrup (HFCS), dextrose or glucose syrup to almost double in price from 2000 – 2010[11] as is shown in Figure 1-2. Also, the recent development of the biodegradable plastic PLA uses corn starch towards its manufacture. If this plastic were manufactured at high volume levels it would be another product that may compete for corn starch use.

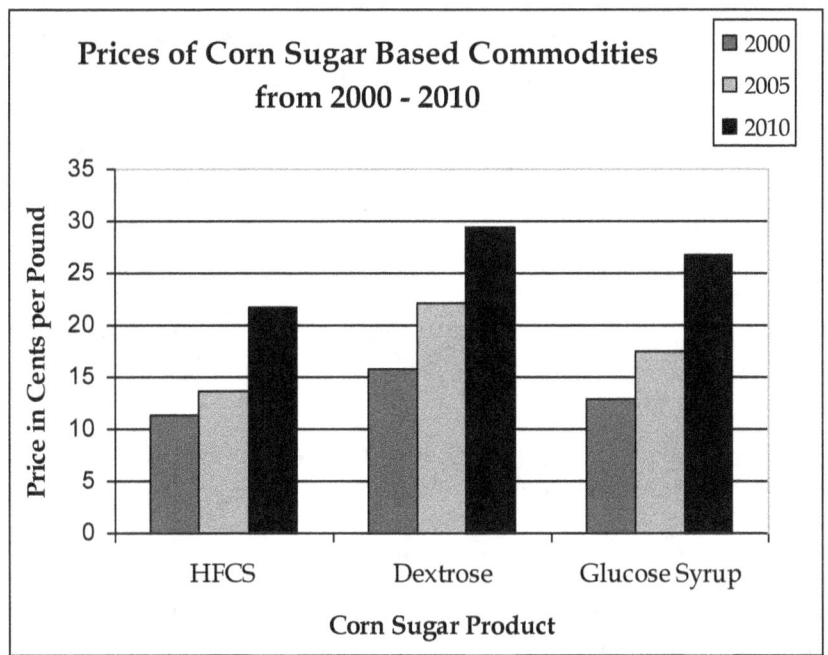

Figure 1-2 : Prices of corn starch related sugar product commodities such as HFCS, dextrose and glucose syrup from the years 2000 – 2010.[10]

Even though other food-energy crops have similar main value components they tend not to favor unhealthy competition between consumable products and energy (biofuel) production. As an example soybean oil crushed from the seed can be converted into biodiesel. However, only a small percentage (around 10 %) of soybean oil is sought for this purpose, whereas the remainder of it becomes fashioned into food or industrial products. In comparison, a much larger percentage of corn starch has been used to manufacture ethanol going back many years. For example, over 27 % of all the corn starch was used to make fuel related ethanol in 1997 when only around 1 billion gallons of ethanol was being manufactured.[12] Nowdays the percentage of corn starch utilized for ethanol manufacture is much larger.

The utilization/production of sugarcane happens to be another food-energy crop worthy of comparison. For example, Brazil manufactures the second largest amount of ethanol per year in the world from sugarcane sources. Taking into consideration the amount of natural resources they have as a country, it is felt by the author that this is a legitimate practice for various reasons.

First of all, sugarcane related products manufacture happens to be well balanced between cane sugar and ethanol. The production of sugar oftentimes takes place concurrently with ethanol manufacture. In many circumstances the making of ethanol does not take away from the potential of raw sugar production. For example, the cane juice will be utilized for sugar while sources such as molasses may be directed more towards ethanol production.

In addition, sugarcane cultivation does not take away as much from farmland resources in Brazil as it is grown along its coastal land area. The total amount of farmland in Brazil exceeds 450 million acres, yet only a small percentage of this has been utilized for sugarcane. For example, in 2007 only 19 million acres was allocated towards sugarcane cultivation.[13] On the converse the US has around 400 million acres of farmland yet close to 90 million acres of this amount is set aside for corn cultivation.

1.4 Significant farm and energy resources are needed to cultivate and process the corn into ethanol

In addition, a significant amount of resources are required to both cultivate and process corn into ethanol. For ethanol production, corn requires substantial farmland acreage, water resources, fertilizer needs, and also large amounts of fossil fuel inputs, both towards its cultivation and conversion into ethanol from refineries. Figure 1-3 demonstrates this concept in terms of the energy requirements necessary to both cultivate corn and process it into ethanol in a dry grind refinery. According to USDA based statistics, around one-third (25.6 kBTU/Gall Ethanol) of the total energy needed to produce ethanol gets expended as farm related processing needs, transportation and distribution while the remaining two-thirds (48.7 kBTU/Gall Ethanol) is spent towards its conversion in the refinery due to production related energy requirements.[14] Therefore, the total processing requirements consist of the energy needed for corn cultivation and ethanol conversion in the refinery. After these two production areas are taken into consideration there is only a net energy gain of around 10 % when comparing it to the intrinsic energy value of the ethanol itself. This value is called Net Energy Value (NEV) and is obtained by subtracting corn ethanol's intrinsic energy value from the total processing energy requirements.

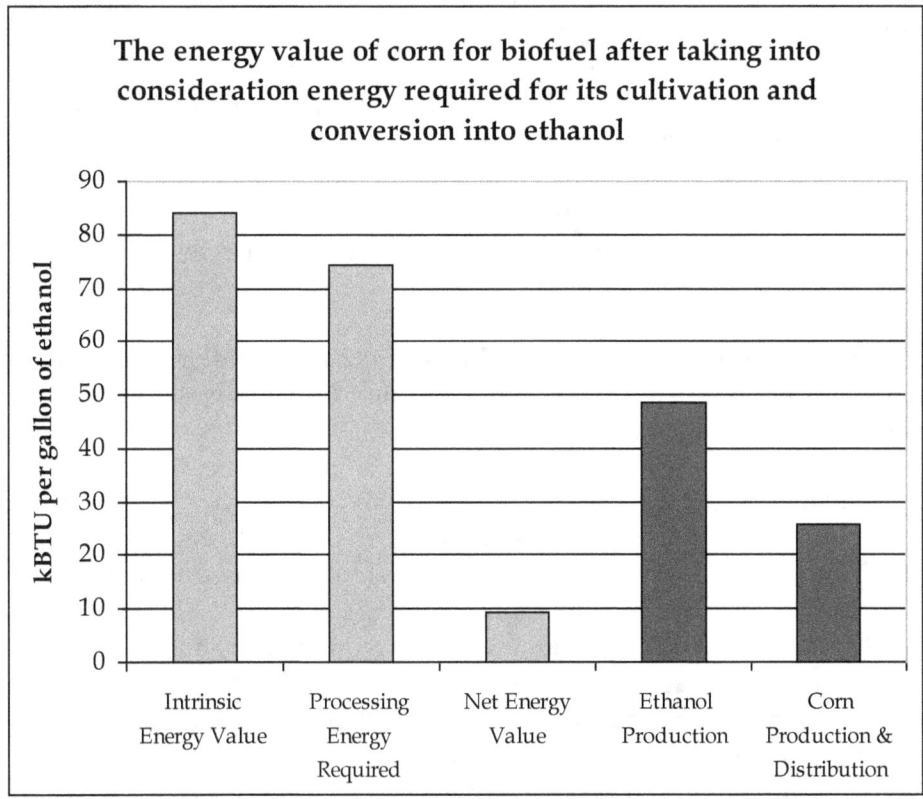

Figure 1-3 : Energy requirements for both corn cultivation and dry grind ethanol processing measured in kBTU per gallon of ethanol produced according to 1996 USDA based government statistics.[13] Net Energy Gain or Value is also estimated to be close to 10 % of what is required to produce the ethanol.

During corn cultivation, sizeable energy requirements involve the manufacture of fertilizer material and the amount of fossil fuel needed to operate farm equipment. Energy needed for fertilizer and fossil fuel energy for machinery make up 40 % and 50 % respectively for the total corn cultivation energy requirement (Corn Production and Distribution value to the far right) as shown in Figure 1-3.[14] Nitrogen fertilizer components consist of the majority (over 80%) of fertilizer needs. Although farmers do not usually manufacture their own fertilizer, the energy required to manufacture it in a fertilizer plant is accounted for when summing this type of production need. Certain disadvantages relate to plants or crops that require high amounts of nitrogen fertilizer. Two such examples have to do with the accompanying soil erosion and water/land pollution. It is estimated that corn uses the most nitrogen and causes the most soil erosion of any food crop.[15]

Water requirements for corn cultivation and ethanol production are also high. It is estimated that at least 15 liters of water are required to produce 1 L of ethanol due to the amount of water that must be initially mixed with the ground corn.[16] Irrigation of corn also requires a large amount of water. Some claims assert that rainwater based irrigation provides for most of this. Regardless, certain areas of the US still need water irrigation brought in from bodies of water or wells for corn cultivation. This includes corn grown in dry, arid, desert regions. In addition, corn raised in desert regions causes both soil erosion and pollution related problems due to its need for irrigation. This not only has adverse effects within the United States but also for other countries that have arid, dry desert regions.

1.5 Production of over 15 billion gallons of ethanol made from corn starch would require significant amounts of farmland primarily set aside for this purpose

As mentioned, corn starch ethanol production currently manufactures just over 13 billion gallons of ethanol per year. This amount approaches the allotted amount of 15 billion gallons of ethanol to be made from corn starch as set by the Renewable Fuel Standards (RFS2). Thirteen billion gallons of ethanol produced from corn starch already requires a significant amount of resources in terms farmland acreage allocation. The production of larger amounts of ethanol per year (i.e., from 15 – 20 billion gallons) would require more substantial resources in terms of farmland acreage.

For example, gasoline containing 15 % ethanol (known as E15) would require at least 20 billion gallons of ethanol produced per year in order to satisfy the blended volume requirement for it. This amounts to another 5 billion gallons of ethanol added to the 15 billion gallons of it already made from corn. Therefore, it would be hoped that this amount be produced from lignocellulosic sources instead of corn. However, if lignocellulosic based ethanol production cannot rise to this amount within the scheduled time period, it is worried that the industry may call for additional amounts of ethanol to be made from corn starch.

The additional amount of farmland required to satisfy these scheduled needs can be projected through calculation. The results are shown in Table 1-4. The estimated farmland acreage required is determined through calculation involving the percentage of animal feed and corn starch ethanol obtained from total corn stock amounts. The yield of corn per acre and the amount of ethanol obtained per bushel are required as well. The main market areas from corn production are shown in Figure 1-4. As shown in the figure, the mass quantity of each market share area is expressed in terms of the number of corn bushels sold for its specific purpose. Corn production statistics for 2010 have shown that ethanol manufacture is 38 % of the total market share while animal feed and residuals accounts for another 40 %.[17] Therefore, both corn starch ethanol and animal feed/residuals represent close to 80 % of the total amount or market share of corn grown, distributed and sold. This amount directly correlates to 80 % of the total amount of land required to make ethanol and animal feed.

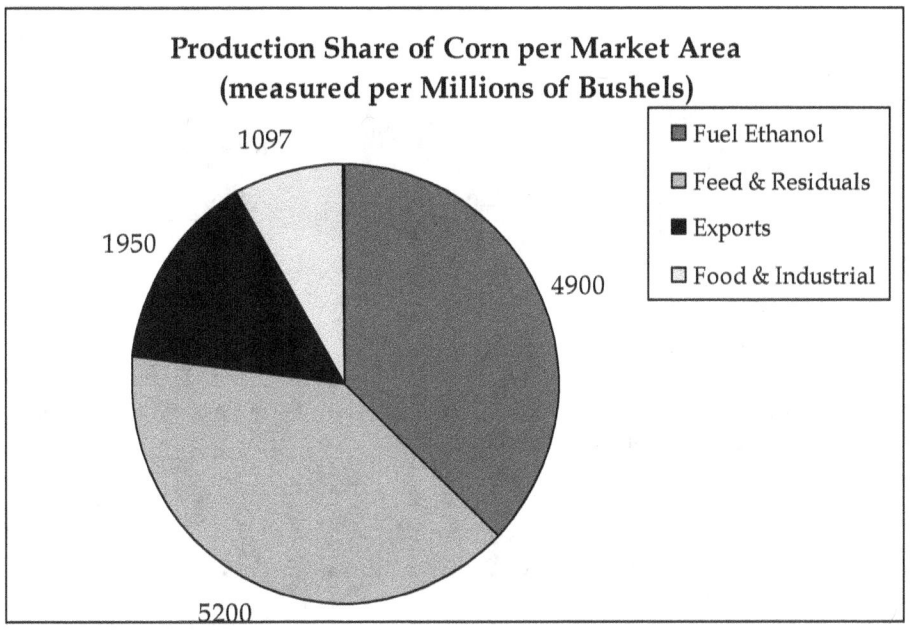

Figure 1-4 : The different production uses of corn as measured by volume amounts in millions of bushels[17]

According to the National Corn Growers Association (NGCA), every bushel of corn can produce 2.8 gallons of ethanol as well as 17 pounds of distillers grains whose primary purpose is for animal feed.[18] Corn starch ethanol production as well as animal feed in the form of Dried Distillers Grains and Solubles (DDGS) are made at the same dry mill refinery location. DDGS represents animal feed obtained by separating the germ, fiber and protein from the starch contained in the corn kernels. This process is roughly depicted in Figure 1-5 below.

In 2010, over 13 billion gallons of ethanol were produced requiring an estimated 69 million acres of farmland. Calculated land acreage amounts for 15 billion and 20 billion gallons of ethanol produced per year were estimated to be 75 million and 98 million acres respectively. Keep in mind that these amounts of land acreage only represent corn grown for animal feed/residuals and ethanol fuel purposes. Additional land requirements that accommodate for the remaining 20 % of corn markets such as exports, food and industrial uses bring the sum total acreage amount to 86 million, 94 million and 123 million acres respectively for ethanol volume amounts of 13 billion, 15 billion and 20 billion gallons per year.

Simultaneous production of DDGS & ethanol from corn

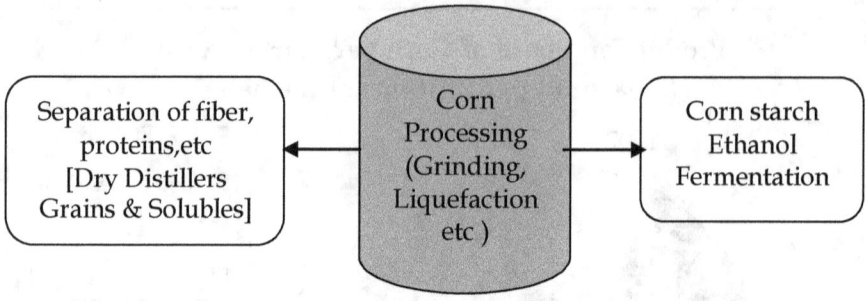

Figure 1-5 : Bioethanol production from the fermentation of corn starch simultaneously produces Dry Distillers Grains and Solubles (DDGS) for animal feed. Such manufacturing processes utilize corn starch for fermentation while the fiber, proteins and germ make up the DDGS.

Projected corn farmland required for current (2010) and 15 or 20 billion gallons of ethanol along with animal feed/residuals

Corn Production Uses	Estimated 2010 land use	Estimated land at 15 billion gallons ethanol	Estimated land at 20 billion gallons ethanol
Ethanol Fuel	33 million acres	36 million acres	47 million acres
Animal Feed	36 million acres	39 million acres	51 million acres
Summed Amount	69 million acres	75 million acres	98 million acres
Total Amount	86 million acres [* actual]	94 million acres [** projected]	123 million acres [** projected]

Table 1-4 : Corn acreage required just to accommodate the simultaneous production of ethanol and animal feed/residuals during 2010 for the projected manufacture of 15 or 20 billion gallons of ethanol.

These calculated results are also based upon an arbitrarily chosen corn yield of 175 bushels of corn per acre. However, in the future higher corn yields per acre may be possible. This would have the effect of using less amount of farmland. Yields between 200 – 300 bushels per acre are expected in the future, but not for another 20 – 40 years.[19] Higher corn yields per acre would then provide for more options towards further food, products and energy applications from corn itself.

1.6 Corn starch ethanol manufacture also adversely affects food supply & prices not limited to the United States itself

The economic success of corn ethanol production depends on several factors that include energy costs (i.e., crude oil prices), overall corn supply and the cost of corn itself. However, it appears that US corn supply has already been adversely affecting world food supply.[6] Recently, this has become a concern because the world commodity food prices have increased dramatically by 60 % since 2009.[20] The USDA states that factors affecting this world price increase include rising energy prices, increased biofuel production and lower world crop yields.[20] Therefore, ethanol biofuel production in itself happens to be a major factor that resultantly affects world food supply and prices.

As stated above, world corn supply also affects world food prices. World corn supply is measured in terms of corn stock to use ratio. This important ratio has been decreasing by noticeable amounts since 2006 and now happens to be at its lowest level since 1995-1996, set at around 4 % of the total yearly supply.[21] The USDA commented that the world corn stock to use ratio is the best indicator of available supply that helps to predict overall corn shortages.[21]

The price ethanol manufactureres pay for corn itself also becomes a significant determining factor as far as recovering operating and capital costs for an ethanol refinery. This can be defined as the relationship between corn purchase costs and ethanol commodity price. The point at where a refinery starts losing profits by paying more for expenses than it makes can be called the 'break even price per bushel of corn'. This estimated price relationship was established by researchers at Iowa between corn cost and ethanol selling price in 2008, at a rate of 50 cents price change per gallon of ethanol for every $1.93 price increase per bushel of corn.[22] Table 1-5 shows the relationship between actual purchase price of corn and the estimated break even price. The corn purchase prices were estimated from the yearly average prices of corn on the stock market.

Chart showing the relationship between the corn purchase price and its break even price

Year	Corn Purchase Price	Break Even Price	+ Price Difference
2007	$3.90 per bushel	$7.48 per bushel	$3.58
2008	$4.60 per bushel	$7.38 per bushel	$2.78
2009	$4.03 per bushel	$3.98 per bushel	-$0.05
2010	$4.23 per bushel	$5.50 per bushel	$1.27
2011	$6.32 per bushel	$6.95 per bushel	$0.63
2012	$7.01 per bushel	$6.48 per bushel	-$0.53
2013	$4.11 per bushel	$6.90 per bushel	$2.79
2014	$3.76 per bushel	$6.36 per bushel	$2.60

Table 1-5 : Break even price relationship between what a corn refinery pays for the purchase of corn and the point at which its purchase exceeds manufacturing costs. These costs consider operating & capital expenses. A larger positive price difference denotes larger profit or recovery of operating costs.[22,23]

The positive price difference of return (thus a positive indicator for profit margin) for corn ethanol refineries has gone through several cycles of rise and fall, similar to what we saw with food commodity crop prices. A few disturbing points are seen from these calculated results. The break even point went below the corn purchase price during 2009 and also 2012. It has also declined in value during the 2010 – 2011 years as well. Only recently (2013 – 2014) have profits gone up again (above $2 per gallon). It appears that major consolidation of the business takes place during the year(s) there is a negative return in profits. This usually means there are less businesses or competitors in this area while fewer companies make larger profits on the ethanol after the dust has settled from a bad year. In the period of 2008-2009, several major corn ethanol producers filed for bankruptcy including the largest west coast producer, Pacific Ethanol, while in 2008 the largest ethanol producer in the country, VeraSun also went bankrupt.[24] It has been disturbing that small or negative profit returns for ethanol producers have taken place during the years of 2009 – 2012. This becomes an important factor since it can take many years to recover the capital expenses that a refinery incurs upon its construction. Capital expenses cover the cost of equipment needed for the refinery such as heat and electrical, wastewater treatment and other areas. These expenses are included in the cost of production as measured in terms of price cost per gallon of ethanol made.

The short history of corn ethanol based production has been marked with financial difficulties related to the continuing rise in the corn prices, the cost of its production, fluctuations in ethanol prices and supply & demand problems.

These problems in turn are related to a shortage and oversupply of corn, bankruptcy issues and lower profit margins for ethanol producers as compared to farmers.

During the first several years of ethanol production it is felt that ethanol producers made their best profit margins (i.e., 2005 - 2006).[25] This was due to the low price of corn. Unfortunately this situation caused an over allocation of corn based farmland resulting in an oversupply of the corn at the expense of other crops. In other words there was an oversupply of corn due to the limited number of refineries built during this time period. However, shortly after this period the number of ethanol refineries increased in number to the point where neither they nor the corn were the limiting factors in the production of corn starch based ethanol (end of 2006).[25]

Our oversupply issues with corn during this period also became a major issue for our neighboring countries as well. During the initial US increase in ethanol production, there were riots in Mexico over the prices of tortillas that immediately raised the concern of the importance of fuel over food production.[26] Others have also attributed the underselling of corn to Mexico and Canada as the cause of job losses to many Mexican farmers[27] and a trade related anti dumping and countervailing trade duties on corn being sold to Canada (latter part of 2005) from the US.[28]

Towards the end of this period (i.e., ~2006), the lack of ethanol refineries were no longer limiting corn starch ethanol manufacture as enough of them were being constructed to fill the fuel demand. Therefore, corn cultivation could not keep up with the potential ethanol volume capacity in terms of the sum total of all the dry grind mill refineries put together. This situation occurred during the period of 2007 - 2008 and caused a corn supply shortage. Ethanol producers were still making profits, but much less of it, since the price of corn doubled in price.[25] The farmers were the ones making a larger relative profit from the sale of the corn during and after this period.

During the last several years (2009 - 2012), it has been difficult for both the farmer and the ethanol producer in terms of production cost and shortage of corn supply. During 2009, the break even price of corn went below profit margin causing bankruptcy for some producers and also resulted in a further consolidation of the business itself.[29] Also, during these years the price of oil rose. This situation brought forth a consequent rise in production costs for both the farmer and the ethanol producer. High oil prices have also indirectly contributed to high food prices due to the cost of processing and crop inputs. Therefore, high oil prices affect the cost to operate farm equipment and machinery as well as the application of fertilizer.

Within the last few years several international organizations have published a study that determined, in their opinion, that corn ethanol production has been partially responsible for causing the current world food problem.[6] One major factor they believe that contributes to this world food problem involves a government mandate establishing biofuel production. In addition, these organizations have recommended that other regulations need to be enforced by governments such as ours in order to ameliorate or fix these problems. Table 1-6 below describes the recommended actions that should take place.[6]

For example, it was recommended that governments exercising a biofuel mandate should try to remove such a policy. They should eliminate any subsidies, payments or tax breaks they believe might be causing food related supply problems. Tax breaks for fuel producers were removed in January 2012. This type of government aid basically gave over $6 billion in tax rebates in 2011 for fuel producers mixing ethanol into gasoline.[30]

Policy adjustments recommended by WTO & other organizations related directly for countries producing biofuels by government mandate	
Remove government based subsidies, tax breaks and payments related to mandated biofuel production (Tax excise break of 45 cents per gallon for fuel producers removed in Jan 2012)	Remove government related biofuel mandates themselves (Relates to producing over 31 billion gallons ethanol by 2022)
Remove foreign trade tariffs on sale of ethanol or other non-corn related feedstocks (US puts a 54 cent per gallon tariff on ethanol sold from other countries)	Utilize feedstocks not related to food and feed markets for ethanol manufacture (One of the subjects covered in this book)
Have biofuel production be more energy efficient and reduce carbon dioxide emissions & resource usage (One of the subjects covered in this book)	Adjust for correcting food commodity prices if supply gets too low or high, induce competition for fertilizer production in general

Table 1-6 : These are some of the recommended actions that organizations such as WTO, FAO, etc recommend to take for governments exercising a fuel mandate on biofuel production.[6] Some of these policies are also some of the main subject material covered in this publication.

As stipulated by these organizations, they do not necessarily have the opinion that biofuel production should be ceased, but instead other types of feedstocks be utilized that do not affect the overall food or animal feed supply. Policy improvements also include making the biofuel production process itself more energy efficient, reduce resource usage and carbon dioxide emissions. These subject areas are addressed in chapter 5 [Alternative alcohols production]. Another important policy restriction that should be replaced includes eliminating trade tariffs on ethanol or biofuel related feedstocks arriving from other countries. The tariff currently involves a 54 cents per gallon placed on the sale of ethanol from countries outside of the United States.

The question then arises as to how we should produce biofuels such as ethanol without utilizing food crops such as corn. Lignocellulosic feedstock sources are the most logical choice that can be implemented towards making biofuel like ethanol. However, even the most apparent method of making ethanol from lignocellulosics has disadvantages related to its production when it is compared to corn starch ethanol manufacture.

1.7 Other alternative ethanol production methods would be preferable to implement over straight lignocellulosic ethanol fermentation

As per US government fuel mandate, it is expected that 31 billion gallons of ethanol be manufactured by the year 2022 from corn starch and lignocellulosic feedstock sources (refer to Table 1-2). It was previously mentioned in this chapter that 15 billion gallons per year of this amount will be produced from corn starch ethanol fermentation. Many then expect that the remaining amount (16 billion gallons per year) be manufactured through the operation of straight lignocellulosic ethanol fermentation based refineries. In other words, lignocellulosic ethanol fermentation is scheduled to become one of the primary sources of ethanol production. However, it has been difficult to establish straight lignocellulosic ethanol fermentation in the recent past. For various reasons this technology has had a hard time competing with corn starch based ethanol manufacture. This is demonstrated by comparing lignocellulosic fermentation refineries to dry grind corn mills based upon their respective production costs and size capacity. This type of comparison helps to establish the case that alternative ethanol production technologies would more effectively produce this additional ethanol rather than relying upon straight lignocellulosic fermentation.

First of all, lignocellulosic fermentation tends to cost more than corn starch ethanol fermentation. The higher production cost per gallon of ethanol made from ligncellulosic fermentation versus corn starch is proof of this assertion. This is mainly due to the overall higher capital cost of building the lignocellulosic refineries when compared to the corn starch types. In 2000 the US government conducted a study that estimated the cost of ethanol production from either a corn starch or a corn stover lignocellulosic fermentative based refinery.[31] Both refineries were estimated to manufacture equivalent amounts of ethanol established at 25 million gallons per year.

Even though the ethanol industry has changed considerably since that time, the comparisons demonstrate the point that lignocellulosic based ethanol fermentation cost more than normal corn starch fermentation until process improvements are made. The comparison in cost of production between corn starch and corn stover ethanol manufacture is shown in Table 1-7 below. As shown in the table, fixed operating costs and the depreciation of capital costs are estimated to be higher in value for lignocellulosic fermentation.

These factors are based upon the purchase and operation (i.e., labor) of the necessary refinery equipment. They lead to an overall higher price in production for lignocellulosic fermentation versus corn starch ethanol manufacture during that time period ($0.88 per gallon versus $1.50 per gallon – year 2000). Variable operating costs include the price of the chemicals, enzymes, denaturant, electricity and wastewater treatment. Fixed operating costs include the amount for labor, supplies and overhead. Capital costs include the price for the essential equipment that often include tanks, heat exchangers and distillation columns as well as equipment for feedstock handling, wastewater treatment, separation and drying and saccharification/fermentation processes. The depreciation of capital cost, defined in this study, is basically the recovery of the capital investment in equipment taken over a 10 year period. So basically one tenth of the overall capital investment is factored into the cost of producing the ethanol per gallon for these refineries.

As reflected in Table 1-7, the overall cost of equipment needed for a lignocellulosic ethanol refinery is almost five times higher than that required for the counterpart corn starch ethanol refinery. Even though these comparisons are over a decade old, the trend towards purchasing more expensive equipment for lignocellulosic fermentation refineries is just as typical today. However, keep in mind that the cost of corn starch production may be somewhat misrepresented in Table 1-7 due to the fact that in the past few years the price for a bushel of corn has gone up from $2 to $5-6 per bushel.

The lower price of corn starch ethanol per gallon is also due to the additional production credits that offset the price of production, which are much higher for corn starch ethanol manufacture. These credits include the sale of dry distillers grains from corn starch ethanol production. However, lignocellulosic fermentation also has credits in the form of electrical energy generation from the refinery.

It should also be stated that corn starch ethanol production has succeeded due to several government monetary assistance programs that have come about during the last several years. So far lignocellulosic ethanol fermentation hasn't received these types of funding breaks. These funding breaks include excise tax credits, producer credits, direct producer payments, commodity credits that offset the purchase price of the corn, direct grants and loans, tax exemptions, income tax credits, etc.[32]

Production cost comparison between corn-starch & lignocellulosic ethanol production

Types of Costs	Dry Grind Corn Ethanol	Lignocellulosic Ethanol
Feedstock Cost	$0.68 per gallon	$0.54 per gallon
Fixed Operating Costs	$0.13 per gallon	$0.36 per gallon
Variable Operating Costs	$0.25 per gallon	$0.17 per gallon
Depreciation of Capital	$0.11 per gallon	$0.54 per gallon
Producer Credits	$0.29 per gallon*	$0.11 per gallon**
Totals	$0.88 per gallon	$1.50 per gallon

* Credit includes sales of Dry Distillers Grains & Solubles
** Credit includes Electrical generation credit to power company

Table 1-7 : Comparison between estimated associated costs of production between corn ethanol and lignocellulosic ethanol[31]

Corn starch ethanol fermentation has also done better in part due to a higher plant volume manufacturing capacity. During this last decade, the capacity of corn ethanol refineries has increased in size so that many refineries produce well over 100 million gallons of ethanol per year. Larger volume ethanol production in general has been known to reduce the cost of production and the recovery of investment and capital costs. However, the trend appears to be that lignocellulosic fermentation refinery capital costs are much higher compared to dry grind mill capital costs as was demonstrated in Table 1-7. For this reason these higher equipment costs may deter the building of larger volume capacity fermentative lignocellulosic ethanol refineries. Evidence of this is shown by examining the size capacity of lignocellulosic fermentative refineries in operation as based upon 2009 statistics shown in Table 1-8.

The majority of the refineries listed in the table use fermentation based manufacturing while only a few implement other thermochemically related technologies. Many of these refineries (i.e., 66 %) produce less than 10 million gallons of ethanol per year. Therefore, most of these refineries do not even meet the 25 million gallons per year DOE base case for the corn stover refinery shown in Table 1-7.

Planned, pilot or commercial lignocellulosic ethanol refineries as categorized by volume production capacity

Production Capacity	Number of Refineries
1 million or less gallons per year	7
>1 - 10 million gallons per year	16
>10 - 30 million gallons per year	8
> 30 million gallons per year	2
Total Number	**33**

Table 1-8 : Breakdown of lignocellulosic ethanol volume capacity for refineries according to projected or actual amounts manufactured (from 2009 data)[33]

Even if ethanol can be produced through technologies that are competitive to corn starch based production there are still questions regarding its sensible application to our current vehicle fuel infrastructure. The next few sections bring up issues questioning whether ethanol should be our preferable choice of biofuel in the future.

1.8 It would be difficult to justify the establishment of an E85 fuel infrastructure on a national level unless regional, localized alcohol refineries were constructed

Flex Fuel vehicles are built to support a higher ethanol blend in gasoline implemented as the fuel known as E85. It is recommended that higher blends of ethanol not be put into regular gasoline spark ignition engines (i.e., non FFVs) due to the corrosive nature of ethanol. In fact, it is uncertain what the vehicle performance and fuel equipment materials compatibility is like with higher blends of ethanol other than E10 in regular spark ignition engines.[34] Therefore, the fuel lines, tanks and other engine components have to be made of different materials in Flex Fuel vehicles in order to support a higher ethanol content in the fuel.

The implementation of an E85 fuel infrastructure has been difficult to establish and maintain during these past years for various reasons. For one thing, our fuel infrastructure is unable to support ethanol transport through our current existing pipeline system due to ethanol's corrosive and hygroscopic (taking in and retaining moisture) nature.[35] Therefore, unless modifications are made to our current pipeline system (cited by example)[36], it would be difficult to transport the ethanol in petroleum based pipelines. Thus ethanol is mainly transported by rail, barge or truck which increases its cost as a biofuel alternative.

The US government conducted feasibility studies as stipulated by the *Biofuels Research and Development* section of the *Energy Independence and Security Act of 2007* regarding ethanol pipeline transportation and the assessment of underground storage tanks in order to accommodate E85 usage as shown in Figure 1-6.[37] As emphasized above, ethanol happens to be incompatible with our existing petroleum based pipeline infrastructure unless the pipelines are cleaned and additional equipment is added to them.[36] In addition, there are complications associated with trying to implement E85 based storage tanks. At fuel filling stations, special E85 type storage tanks would need to ordered and placed underground since the existing fuel storage tanks are not adequate for this purpose. This type of infrastructure change makes it more difficult to implement E85 at regular fuel filling stations.

Ethanol pipeline transportation & storage tank assessments as done under the 2007 EISA

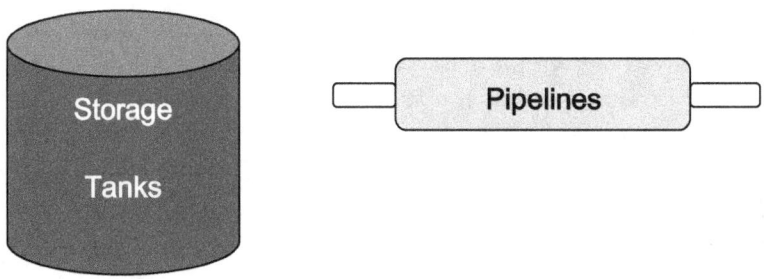

Figure 1-6 : Ethanol pipeline transportation feasibility and underground storage tanks assessments were done according to the Energy Independence and Security Act of 2007 [EISA].[43]

For reasons such as these, it would be difficult to support E85 due to its incompatibility with our fuel transportation and dispensing infrastructure. The inability to transport ethanol cost effectively at far distances have limited the majority of E85 infrastructure and use to the Midwestern states.[26] Table 1-9 lists the disadvantages of trying to implement an E85 fuel infrastructure.

Another disadvantage regarding higher ethanol fuel blends such as E85 include obtaining less gas mileage when compared to driving one's vehicle with gasoline. This is due to the fact that ethanol has a much lower energy density (i.e., energy content) than gasoline. It has been shown that E85 provides approximately 30 % less gas mileage as opposed to gasoline. It has also been expressed that it would be difficult to construct enough E85 fuel stations required to keep up with the fuel demand correlated to the number of FFVs traveling on our roads and highways.[34] A possible explanation for this may be due to the cost of underground storage tanks that are reported to cost around $100,000 due to the special lining they require.[26]

Disadvantages of supporting an E85 infrastructure

Disadvantages
Difficult to build enough E85 stations to satisfy demand
Difficult to transport ethanol with Pipelines
Ethanol based storage tanks can be expensive & difficult to implement
E85 has lower fuel mileage as compared to gasoline

Table 1-9 : The disadvantages of trying to implement an E85 fuel infrastructure for Flex Fuel Vehicles (FFVs)

1.9 Hydrous ethanol is a cheaper form of ethanol that may help solve its transportation issue as well as enhance fuel performance

Past research demonstrates that hydrous ethanol (water mixed in) instead of the anhydrous version maybe easier to incorporate into our fuel infrastructure.[38,39] Ethanol sold as the fuel additive is anhydrous meaning that it contains greater than 99 % ethanol. Hydrous ethanol is a high mixture of ethanol (> 90 %) with small amounts of water (< 10 %). It is also known that hydrous ethanol tends to be much cheaper to manufacture since it does not require a further dehydration step. This is demonstrated in Figure 12 below.

It is claimed that hydrous ethanol/gasoline has been shown to yield better engine performance when compared to its anhydrous ethanol / gasoline counterpart fuel mixture.[40] These claims assert that some water mixed in with the gasoline/ethanol fuel tends to increase compression ratio, heat of vaporization as well as lowers the engine operating temperature.[40]

Hydrous ethanol may be more suited for transportation in pipelines as well. For instance, Brazil successfully transports their ethanol through pipelines but they do this as the combination of both hydrous and anhydrous ethanol shipped together across the pipelines.[38] In other words, both types of ethanol are manufactured and transported together in a consecutive manner. Important considerations of such an infrastructure would include the percentage of ethanol refineries that ship ethanol by pipeline, how recently built is the pipeline infrastructure, the cost of the pipeline itself including materials improvements and the average mileage of pipeline per shipment.

Manufacturing processes required to make anhydrous ethanol

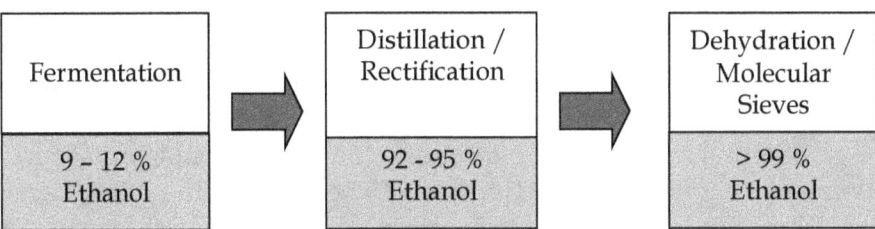

Figure 1-7 : The manufacture of hydrous ethanol can eliminate the need to use a further dehydration step shown in the far right of the figure. The elimination of a dehydration step also would have the tendency to lower manufacturing costs.

1.10 A greater understanding of biofuel concepts should be required when preparing/proposing further government based policies and legislation

In the future it is proposed that our existing corn starch refinery technology be retrofitted to accommodate the manufacture of lignocellulosic ethanol. In theory, the basic operational equipment contained in a dry grind corn mill ethanol refinery could be modified for this purpose. This type of policy has been stipulated in the *Biorefinery Energy Efficiency* section of the *Energy Independence and Security Act of 2007*.[41] However, this policy may be difficult to carry out for a number of reasons. First of all, basic inherent differences exist between the fermentable and non-fermentable components of corn kernels and lignocellulosic materials. Corn contains large amounts of proteins and oils whereas lignocellulosic materials usually do not. In addition, lignocellulosic materials contain lignins which are not found in corn kernels. Corn also contains starch based polysaccharides that are easier to break down than the cellulose or hemicellulose materials from lignocellulosic sources. Lignocellulosic fermentation also requires fermentation through different types of enzymes and yeast cells. Therefore, retrofitting existing corn starch refineries may be more difficult than it appears. These facts should have been apparent to those who set forth legislative policies regarding biorefinery utilization.

It is evident that our recent past government based policies such as fuel mandates and projected ethanol based infrastructure changes have had some unforeseen complications. The international community has the opinion that a fuel mandate has caused economic problems with food supply and prices due to ethanol manufacture from corn. Past energy acts were also inaccurate at assessing the accommodation of ethanol as a biofuel. The goal of such legislation was to ascertain whether such infrastructure changes were practical and cost effective. It becomes fairly evident that policies such as transportation through existing pipelines, installation of underground tanks and feasibility of retrofitting of corn starch ethanol refineries are not prudent to carry out. However, these complications could have been avoided by providing the correct information to government officials. Therefore, it is felt that a greater understanding behind the science and manufacturing processes for alternative biofuel manufacture should be essential for future scientists and policy makers as well as for the general public. This type of information is essential to the successful implementation of practical, effective and efficient technologies for future renewable biofuels production.

Biofuel production in general is still in its infancy. So far, our technology has primarily produced just the biofuels of biodiesel and ethanol on a large production scale. However, the fact is that there are considerably more vehicle fuel types that can be made from certain types of biofuel manufacture. For example, the same type of hydrocarbons that make up gasoline and diesel can be made from renewable resources as well. In reality, biofuels manufacture mainly requires a set of hybrid manufacturing methods that combine initial biomass processing with existing or improved refinement technologies that already exist.

The main question that people in general have is whether biofuels manufacture will succeed since many feel that it has to be competitive in price and application when compared to petroleum based vehicle fuels. The answer to this should be a resounding YES. However, the key to this realization involves a combination of factors that include a new philosophy in constructing and utilizing biorefineries as well as manufacturing additional useful byproducts made during the biofuel manufacturing process. These additional saleable products have the potential to become the main factor leading to the future success of biofuel production. Similar to the many petrochemicals in use, synthesized chemicals or biochemicals made as byproducts from biofuels manufacture can serve the same role in satisfying our material needs. The non-believer should disregard the hearsay that the same or equivalent types of chemicals cannot be made from renewables. It will only require further research and application to make these same types of chemicals a reality. Therefore, the remaining material in this book attempts to demonstrate this and all the aforementioned points through concrete examples of production technologies involving many types of biofuels as well as the material byproducts made with them. In addition, the sensible application of renewable resources utilized towards these production goals is further addressed and discussed throughout the book.

REFERENCES:

1. **Historical US fuel ethanol production**, Renewable Fuels Association – http://www.ethanolrfa.org/
2. **Renewable Fuel Standards (RFS2) – 2010 and beyond** – National Ethanol Conference 2010 – Renewable Fuels Association
3. **Breaking the biological barriers to cellulosic ethanol : A joint research agenda** – Office of Energy Efficiency & Renewable Energy – Introduction pg. 4 by J. Houghton, S. Weatherwax, J. Ferrell [Jun 2006]
4. **Corn acres expected to soar in 2007, USDA says** – Mar 2007, USDA National Agricultural Statistics Service
5. **USDA Crop Values 2014 Summary** – USDA National Agricultural Statistics Service [Feb 2015]
6. **Price volatility in Food and Agricultural Markets : Policy Responses** – by the FAO, FAD, IMF, OECD, UNCTAD, WFP, the World Bank, the WTO, IFPRI, IN HCTF [June 2011]
7. **Why ethanol production will drive food prices even higher** – http://www.earth-policy.org/plan_b_updates/2008/update69 -- by LR Brown [2008] -- accessed May 2014
8. **2012 World of Corn** by the National Corn Growers Association, http://www.ncga.com/worldofcorn -- accessed May 2014
9. **Crop production 2011 summary** – by the USDA National Agricultural Statistics Service [Jan 2012]
10. **Opportunities for renewable bioenergy using microorganisms** – Biotechnology and Bioengineering Vol 100 No 2 pgs 203 – 212 by B. Rittmann [2008]
11. **World & US Sugar and Corn Sweetner Prices** – Tables 7,8,9 US Prices for Glucose Syrup, Dextrose and HFCS – by the USDA Economics Research Service
12. **Ethanol, citric acid and lactic acid use of corn as a feedstock** – Industrial uses of Agricultural Situation and Outlook Report – 1998
13. **Brazil's ethanol industry** – by D. Hofstrand – AgDM Newsletter [Jan 2009]
14. **The energy balance of corn ethanol : An update** - USDA by H. Shapouri, J. Duffield, M. Wang [2001]
15. **NAS Frontiers in Agricultural Research, Food, Health, Environment and Communities**
16. **Biofuel impacts on world food supply : use of fossil fuel, land and water resources** – Energies Vol 1 No 2 pgs 41 – 78 by D. Pimentel, A. Marklein, MA Toth et al [2008]

17. **2011 World of Corn** – by the National Corn Growers Association – http://www.ncga.com/worldofcorn -- accessed May 2014
18. **Corn distillers grain value added feed for beef, dairy beef, dairy, poultry, swine, sheep** – Sept 2008 by the National Corn Growers Association
19. **Biomass as a feedstock for a bioenergy & bioproducts industry : The technical feasibility of a billion ton annual supply – Factors increasing biomass resources from agriculture** by US Department of Energy and US Department of Agriculture – R. Perlack, L. Wright, A. Turhollow, R. Graham, B. Stokes, D. Erbach [Apr 2005]
20. **Why another food commodity price spike?** – by AmberWaves [Sept 2011] – http://www.ers.usda.gov/AmberWaves/.. – accessed May 2014
21. **Feed Outlook – US and global corn stocks projected lower** – ERS USDA [Feb 2011] by E. Allen, H. Lutman, T. Capeheart
22. **How low will corn prices go?** – Iowa Ag Review Vol 14 No 4 [Fall 2008] by B. Babcock
23. **Corn and Ethanol monthly commodity futures price charts** – http://futures.tradingcharts.com/ -
24. **Pacific Ethanol units joins others in bankruptcy court** – by M. Andrejczak [May 2009] – http://www.marketwatch.com/story/pacific-ethanol-joins-others-in-bankruptcy-court -- accessed May 2014
25. **Who profits from the corn ethanol boom?** – Ag Decision Maker Newsletter [Sept 2008] by D. Hofstrand – Iowa State University Extension and Outreach
26. **Growing America's fuel : an analysis of corn and cellulosic ethanol feasibility in the United States** – Clean Technologies and Environmental Policies Vol 12 pgs 373 – 388 by D. Somma, H. Lobkowicz, JP Deason [2010]
27. **How the (finally ended) corn ethanol subsidy made us fatter** – [Jan 2012] by B. Watson – http://www.dailyfinance.com/.. – accessed May 2014
28. **Unites States requests WTO consultations with Canada over duties on grain corn** – [March 2006] Office of the United States Trade Representative – http://www.ustr.gov/-
29. **Crystal Eth: America's crippling addiction to taxpayer financed ethanol** – The Institute of Energy and the Environment (Vermont Law School) [Apr 2011]
30. **Record food prices linked to biofuels** – by K. Bullis [June 2011] – http://www.technologyreview.com/energy/37848
31. **Market penetration of ethanol** – Renewable and Sustainable Energy Reviews Vol 14 No 1 pgs 394 – 403 by KR Szulczyk, BA McCarl, G. Cornforth [2010]

32. **Ethanol fuel incentives applied in the US – Reviewed from California's perspective** – California Energy Commission [Jan 2004] by T. MacDonald
33. **Development and status of dedicated energy crops in the United States –** In Vitro Cellular and Developmental Biology (Plant) Vol 45 pgs 282 – 290 by RW Jessup [2009]
34. **National biofuels action plan –** Biomass Research and Development Board [2008] http://www1.eere.energy.gov/biomass/pdfs/nbap.pdf
35. **Metabolic engineering of microorganisms for biofuels production : from bugs to synthetic biology to fuels** – Current Opinion in Biotechnology Vol 19 pgs 556 – 563 by SK Lee, H. Chou, TS Ham, TS Lee, JD Keasling [2008]
36. **Kinder Morgan – central florida pipeline ethanol project** – http://www.afdc.energy.gov/pdfs/km_cfpl_ethanol_pipeline_fact_sheet.pdf
37. **Ethanol pipeline feasibility study (section 243) & Renewable fuel dispenser requirements (section 242)** – Energy Independence and Security Act of 2007
38. **Liquid Transportation Fuels for Coal and Biomass : Technological Status, Costs and Environmental Impacts** [2009]
39. **Testing the water** – http://www.ethanolproducer.com/articles/3981/testing-the-water – accessed May 2014
40. **Perspective : US needs to transition to hydrous ethanol as the primary renewable transportation fuel** – by BJ Donovan [Aug 2009] – http://www.greencarcongress.com/2009/08/donovan-hydrous-20090830.html -- accessed May 2014
41. **Biorefinery Energy Efficiency – Section 224** – Energy Independence and Security Act of 2007

Chapter 2

The Emerging Biofuel Infrastructure

2.1 Factors that would contribute to an improved biofuel infrastructure

Ethanol has already established a biofuel infrastructure centered upon the production and processing of starch from corn. However, the resulting increase in corn production has the tendency to compete disfavorably for our farmland resources. It would be best if other systems were set in place that would produce biofuel without adversely affecting our valuable farm resources. In fact, establishing several types of biofuel production systems based upon non-food renewable feedstocks would help to conserve many of these resources. These types of systems can come forth by bringing together the proper technologies and methods associated with alternative biofuels manufacture. Improved manufacturing would then help spark the further development of a more modern biofuel infrastructure.

One of the most important concepts being developed that can assist towards establishing a biofuel infrastructure would be the ability to affordably produce biofuel from renewable sources while making additional chemical byproducts that compete with petrochemicals. Many of the pieces for such an infrastructure are already being developed that will help to accomplish this feat. For example, a new trend in producing biofuel at the refinery level is coming forth. It is known as the establishment of the integrated biorefinery.

The integrated biorefinery has the potential to make biofuel affordably while manufacturing alternative byproducts at the same time. Imagine how such systems would favorably compete with petroleum based fuel as well as the petrochemicals and products made from petroleum. An increased revenue stream would invariably result from the implementation of integrated biorefineries. This book gives several examples of integrated biorefineries. Definitely, the new manufacturing markets associated with biochemicals or biobased byproducts would be established from the practice of utilizing integrated biorefineries for biofuel production.

Another idea that could help further the establishment of a biofuel infrastructure would be the production of renewable based alternative organic compounds that serve as fuel additives or fuel amendments. These compounds have the ability to enhance the octane value of a gasoline fuel. Compounds known as oxygenates are a class of organic compounds that can serve as fuel additives. The use of these oxygenates brings forth options associated with replacing ethanol as the fuel additive of choice. These oxygenates are made from synthesis methods involving the combination of other chemicals that can also be derived from renewable sources. Therefore, a biofuel infrastructure in itself could theoretically flourish from the manufacture of these oxygenates from a variety of renewable sources. Alternatively (or concurrently) an included gasoline fraction or amendment can be made from straight hydrocarbons obtained from renewables. These compounds from renewable sources are already similar in chemical composition to what is contained in gasoline fuel.

The manufacture of an alternative set of alcohol based fuels from renewables is another option toward the expansion of a biofuel infrastructure. These alcohols include those other than just ethanol. Such a mixed set of alcohols contains ethanol as well as higher molecular weight alcohols, all of which would make an excellent fuel amendment to gasoline. Higher molecular alcohols can also be made singularly. The alcohol isobutanol is such an example. A set of individualized alcohols would also be helpful in producing other fuel additive oxygenated organic compounds. These alcohols are useful since the synthesis of oxygenates oftentimes requires their addition to another organic compound.

The development of an alternative fuel source would also serve to help establish a biofuel infrastructure. Renewable diesel is such a source. A comparison between renewable diesel and gasoline shows that renewable diesel is actually the more practical and easier type of fuel to produce. This is due to its wider flexibility in the inclusion of organic compound types and wider carbon number size range for the fuel compounds. Large amounts of renewable diesel have the potential to be made from a number of renewable sources.

Another reason for producing further biofuel is the advent of more advanced technologies that have the ability to utilize such fuels. The recent development of an alternative engine is such an example. This type of engine is a hybrid between spark ignition and compression ignition (diesel) engines. An engine such as this may be more suited to operate on biofuel rather than our established gasoline based fuel.

The success of establishing biofuel manufacturing also depends upon utilizing the appropriate renewable feedstocks. Feedstocks such as energy crops, crop residues and algae are excellent alternatives to corn that have the capability of producing large amounts of biofuel. Section 2.7 describes how some of these renewable feedstocks can be better implemented towards making the biofuel. There is no reason why biofuel itself cannot be made from renewable wastes that emanate every year from various sources. Alltogether there are many reasons for establishing a dependable biofuel infrastructure. Most importantly, the realization of such an infrastructure would, in general, increase our country's manufacturing capability.

2.2 Renewable organic hydrocarbons and oxygenated compounds can enhance the engine performance and octane value of gasoline.

Oxygenated fuel additives improve engine combustion and give gasoline a higher octane value as well as help to reduce exhaust emissions. For a considerable time, lead was added to gasoline as an octane enhancer. It was phased out when its adverse pollutant effects were declared to be detrimental to human health. The ether compound known as MTBE was then added to gasoline to fulfill the mandated EPA oxygenated fuel requirement in reformulated gasoline. However, it was banned as a fuel additive by many states in 2004 due to the fact that MTBE contaminated soil and underground water sources.[42] Subsequently ethanol was chosen to replace MTBE as the fuel additive necessary to fulfill oxygenated fuel requirements and assist towards enhancing the octane value of reformulated gasoline.

Single cylinder engine modification and conditions used to determine octane numbers

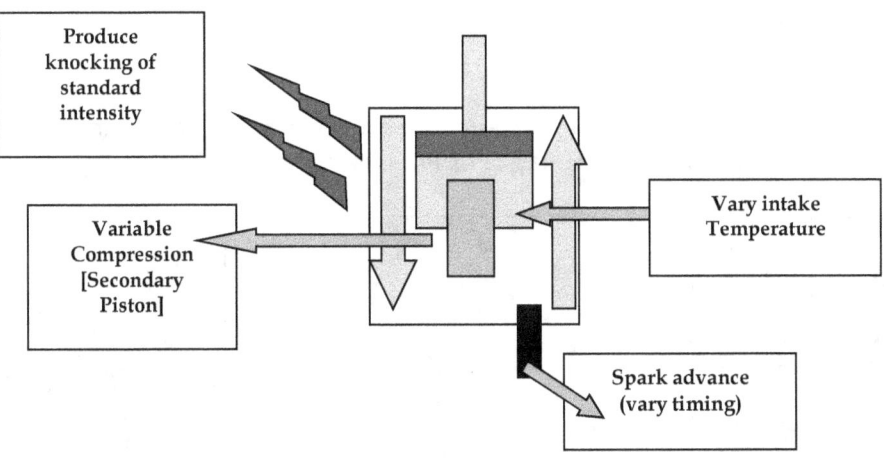

Figure 2-1 : The various engine testing conditions implemented on single cylinder modified engines useful for the determination of research octane number or motor octane number. Engines can be modified to contain a secondary piston that has the ability to vary compression ratio. Tests are conducted in such a way as to produce a knocking of standard intensity from a test fuel that is compared to a reference octane based fuel.

In order to appreciate the importance of the octane value in gasoline as the determining factor for fuel performance in spark ignition engines, it is necessary to understand how the octane number is measured in a fuel and what types of compounds contribute to its overall value. The octane number of a given fuel is determined through various single cylinder modified engine tests as stipulated by specific ASTM testing methods.[43] Experimental conditions within an engine cylinder that determine octane number are intake temperature, spark ignition timing, engine speed and compression ratio as are shown in Figure 2-1. During these engine tests, a standard reference fuel and a sample test fuel are operated alternatively in the engine until they produce a knocking of standard intensity. After testing, both fuels have their knock of standard intensity compared to one another in order to help determine the specific octane number of the sample fuel being tested.

It's also important to understand how a reference fuel helps determine the octane number of a sample fuel based upon its composition. The reference fuel consists of two specific compounds whose octane ratings have already been established. The first compound, n-heptane, has an octane value of zero (0) while the compound iso-octane has been assigned the octane value of 100. These two compounds are mixed together in a certain ratio in order to make a reference fuel of a specific octane number as is shown in Figure 2-2. Thus, they can be mixed together to produce a reference fuel with any desired octane value. Therefore, the octane value of the reference fuel is primarily determined by the amount of iso-octane placed into it.

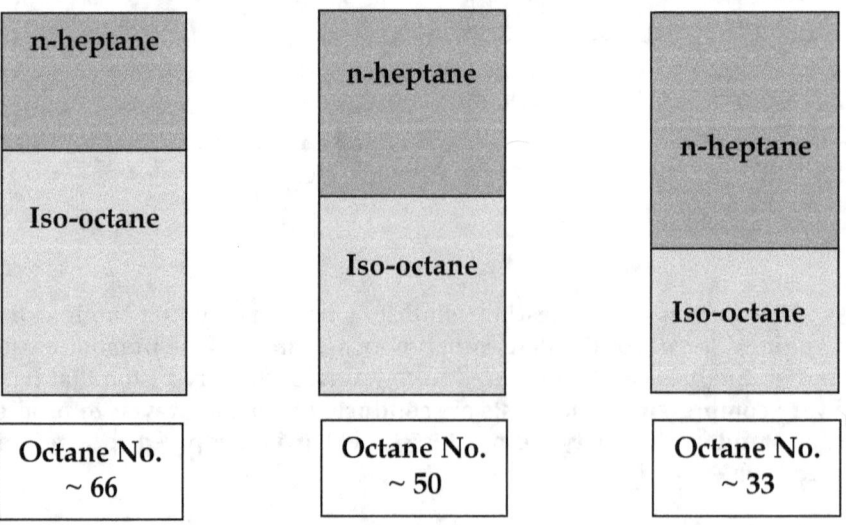

Figure 2-2 : A reference test fuel of any octane number can be made by combining different percentages of iso-octane mixed into n-heptane.

The consistency of gasoline in terms of its reformulated chemical composition primarily determines its octane value. Oxygenated organic compounds such as ethanol added to gasoline can help establish a strong octane value in the fuel. However, it should be known that certain renewable based hydrocarbons can also enhance the octane value of gasoline if they were to be added to it as a fuel amendment. The type of renewable hydrocarbons that impart beneficial octane value would be similar in structure to those already contained in the petroleum based gasoline. Since only certain types of hydrocarbons contained in gasoline give it a high octane value, it's important to know which types they are. In order to grasp this concept one must examine the array of organic hydrocarbons found in petroleum gasoline.

The five major types of organic compounds contained in gasoline are shown in Table 2-1. Three of these, the n-parrafins, cycloparrafins and isoparrafins constitute an organic compound category known as parrafins. Together they make up close to 60 % of gasoline by chemical composition.[44] The parrafins impart the largest range variation in octane value to the gasoline. This is demonstrated by examining the octane range of isoparrafins as compared to n-paraffins. Isoparaffins have octane values that range from low to very high due to variations in the chemical size and structure, while n-parrafins usually have low octane numbers that are primarily determined by its carbon size.

Aromatics, the fourth type, make up another 30 % or more of the total gasoline composition. They normally impart a very high octane value since many of these aromatic compounds have octane numbers over 100. Olefins, the fifth type, are similar to parrafins except that they contain double bonds in their chemical structure that tends to give them a better octane number when compared to their similar paraffin equivalents. Olefins on the average make up around 10 % of the total gasoline composition.[44]

One potential approach towards improving or maintaining the octane value of petroleum gasoline involves reformulating its hydrocarbon composition with high octane hydrocarbon fractions. This could just as well be done with renewable based hydrocarbons. The types of hydrocarbons added would normally be aromatics and/or isoparaffins. However, concerns with aromatics relate to their adverse effect to human health. This especially becomes a concern when they are inhaled by humans if they exit the vehicle as unburned hydrocarbons. Therefore, this section concentrates more on the production of isoparaffins from renewables for octane enhancement rather than aromatics.

Gasoline compound breakdown as stipulated by compound class, percentage and octane range

Compound Class	Octane Range	Percentage	Structure Example
isoparrafins	low to very high	~ 35 %	
n-parrafins	Very low to low	~ 15 %	
Cyclic parrafins	Moderate to high	~ 10 %	
Aromatics	high to very high	~ 30 %	
Olefins	Moderate to high	~ 10 %	

Table 2-1 : The characterization of gasoline by compound classes contained at certain percentage composition that also impart an octane value range to the gasoline fuel formulation itself.[44,45,46]

Renewable based isoparaffins can be produced from refineries that utilize thermochemical gasification technology. These types of biofuel refineries implement catalytic systems that produce initial hydrocarbon cuts in a certain size range. These initial cuts then undergo a fuel upgrading process. This includes changing the composition of this initial biofuel hydrocarbon cut using the following measures.

- Convert larger sized n-paraffins into middle sized distillates such as isoparaffins from the processes of cracking, oligomerization and isomerization.[47,48]
- Increase the proportional amount of isoparrafins having optimal carbon sizes from five to eight carbons through a process known as alkylation.[49]

Besides the renewable refined hydrocarbons produced from thermochemical processing, a large array of oxygenated compounds other than ethanol or MTBE are just as effective in enhancing the octane value of gasoline if they were to be mixed into it. These alternative oxygenated fuel compounds can all be manufactured from renewable resources that do not have to rely on corn production. Therefore, both biobased oxygenated and hydrocarbon fuel compounds have the potential to become more widely incorporated into our existing vehicle fuel production, whether it is gasoline or diesel. They would help improve its chemical as well as its physical properties.

In addition, the widespread manufacture of both hydrocarbon and oxygenated biofuel compounds has the potential to enhance the biofuel industry in further applications. One such application includes the use of biofuel in the recently designed engine known as the homogeneous charge compression ignition (HCCI) engine, a topic discussed in the next section.

2.3 Biofuel compounds are ideal for newer engine designs such as the Homogeneous Charge Compression Ignition (HCCI) engine

Recent automotive research has led to alternative engine designs that have some advantages over the conventional spark ignition or even a diesel combustion engine. One such engine design known as the Homogeneous Charge Compression Ignition (HCCI) engine operates on the principle of igniting a premixed fuel without spark assist. It runs similar to the regular diesel engine since the fuel-air mixture self ignites. However, the manner of introducing the fuel and the resultant timing of ignition differ.

This engine design has gone through extensive experimental research within this last decade.[50,51,52,53] A number of improvements have motivated its design. One has to do with the idea that engine performance can be improved through better combustion efficiency resulting from a controlled mixture of inputted fuel. Due to the homogeneous nature of the vaporous fuel mixture introduced into the HCCI engine cylinder, the fuel-air mixture self-ignites while burning cleaner and more evenly. This type of engine accomplishes this through the tight control or regulation of certain engine inputs & outputs. As shown in Figure 2-3 below, combustion efficiency is improved because ignition can occur at the optimum point of the compression cycle and burns as a uniform mixture instead of inducing flame propagation with uneven burning.

Therefore, a uniform burn of the fuel-air mixture promotes the efficient transfer of energy to the piston from the force of combustion. As previously stated, this takes place when an evenly vaporized and distributed fuel-air mixture enters the cylinder at the right temperature and pressure conditions. The justification for such a design is based on the idea that a complete and proper combustion should allow the vehicle to obtain optimal fuel consumption and operate at a lower temperature. The appropriate engine operating temperature allows it to efficiently transfer the chemical energy from the fuel into mechanical energy driving the piston. On the converse, an inefficient engine tends to waste energy through overheated gases leaving the engine through the exhaust.

The physical nature of the air fuel mixture of conventional combustion and homogeneous charge based combustion

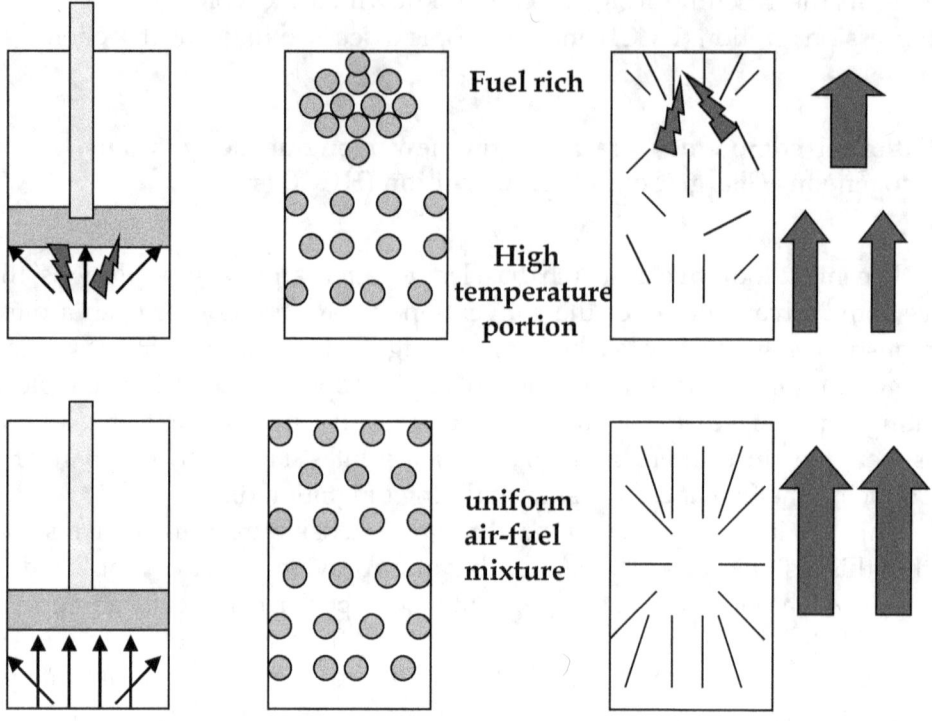

Figure 2-3 : A normal spark ignition combustion engine has a non-uniform air fuel mixture where there are fuel rich and high temperature separate regions that are contained in the engine cylinder that cause it to burn unevenly causing a flame propagation indicative of an inefficient combustion process.

Achieving the right compression ratio and operating conditions favorable to optimum compression in the engine cylinder requires tight control of several input variables. As shown in Figure 2-4, these variables include effective control of the inlet air-fuel mixture temperature, the amount of air to fuel introduced into the cylinder, the timing of the fuel injection and efficient exhaust evacuation.[54] In addition, the HCCI engine can be outfitted to adjust the compression ratio based upon the type of fuel utilized in the engine.

HCCI input parameters that help to determine efficient combustion performance

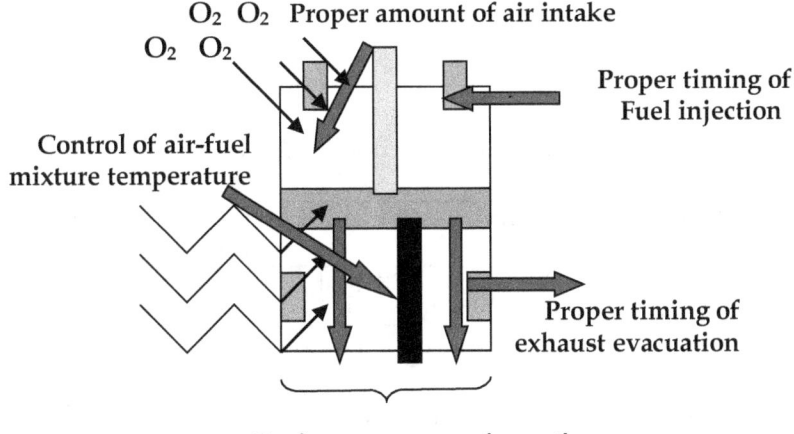

Figure 2-4 : Several input parameters that control the operation of the combustion chamber include proper timing of fuel injection and exhaust evacuation through outlet ports. Other factors include optimum compression ratio, proper amount of air intake and control of in-cylinder temperature.[54]

The development of this engine has been based upon the chemical nature of the fuel input into it. Fuel tests of this engine have utilized a simple set of hydrocarbon or oxygenated fuel compounds. Fuels with simpler chemical composition tend to operate better in such engines. They introduce a more homogenous uniform fuel mixture into the engine cylinder. More chemically complex fuels usually do not reach this type of uniformity in composition when placed into the engine as an air-fuel mixture. Therefore, their performance is not as controllable in engines that have the goal of maintaining a homogeneous air-fuel mixture.

Complex fuels such as gasoline contain hundreds to several thousands of individually distinct hydrocarbons within a given size range. This large compositional variation makes it difficult for the fuel to be introduced into an engine as a homogenous vapor based mixture. Such chemical complexity of the engine fuel becomes an important consideration since very tight controls of engine parameters are required during HCCI engine development, testing and operation. A less complex fuel allows for better control of these engine parameters. Biofuels themselves tend to be less chemically complex than petroleum gasoline or diesel as they usually contain a simple set of organic hydrocarbons or oxygenated compounds. Therefore, certain biofuels are ideal for operation in an HCCI engine. Experimental studies have been executed on HCCI engines using a variety of biofuels, including gaseous based fuels, alcohols, ethers, esters and biodiesel.[56,57,58,59,60,61,62] Table 2-2 contains a list of the specific biofuels tested in HCCI engines either as single based compounds or a combination of two to three simple organic compounds that contain at least one biofuel compound in them.

Some biofuels that have been experimentally tested in HCCI engines

Single biofuel compound	Mixture of two to three single compounds
Hydrogen, ethanol, methane, biodiesel	Ethanol-heptane, butanol-heptane, DME/methanol, DME/methane, butyl-propyl-ethyl acetates

Table 2-2 : Several types of biofuels, most of them oxygenated based, have been tested in HCCI engines either as single compounds or a mixture of two to three single biofuel compounds.[54,55,56,57,58,59,60,61]

The realization of an HCCI engine in the marketplace still requires a substantial amount of research and development since several setbacks have prevented its immediate production. These include the inability of the engine to respond properly to high load conditions such as carrying higher weight, traveling up inclines and the onset of acceleration to do so. In addition, sudden start-stop scenarios like immediate thrust from the engine idle position are yet to be solved as well.

2.4 A variety of oxygenated fuel additives enhance the fuel properties of both gasoline and diesel fuels

Oxygenated biofuel compounds make ideal fuel compounds or fuel additives due to their enhancement of combustion during use. These types of compounds normally contain high octane values. They also contain beneficial chemical properties. Therefore, these fuel compounds do not normally have problems with water miscibility, toxicity, pollution, volatility and corrosion. In reality, a large choice of these oxygenated compounds could effectively serve as fuel compounds and/or additives. The organic compound classes that constitute these oxygenated compounds are shown in Table 2-4. These include ketones, carbonates, ethers, esters and linear or branched alcohols. Many of these compounds have already been tested in engines as fuel additives to gasoline or diesel fuel.[63,64,65,66,67] The compounds can be practically synthesized from renewable biomass in combination with one of many types of alcohols.

Classes of oxygenated organic compounds that serve as fuel additives

Ketones	Carbonates
Esters	Ethers
Linear Alcohols	Branched Alcohols

Table 2-4 : The oxygenated organic compounds of ketones, carbonates, esters, ethers, linear and branched alcohols have been tested and utilized in spark ignition and diesel combustion engines as fuel additives with hydrocarbon based vehicle fuels.[63,64,65,66,67]

Below in Figure 2-5 is a list of synthesis schemes that can be utilized to create some of these oxygenated fuel compounds. The synthesis can take place by modifying an alcohol, combining two alcohols, mixing a natural compound with an alcohol (three examples) or mixing certain organics with glycerine (two examples). The ether ethyl tert butyl ether (ETBE) [1] is made through the reactive distillation of isobutene and ethanol.[68] Isobutene can be obtained through the dehydration of the alcohol isobutanol. Methyl ethyl ketone [2] is produced from the conversion of the alcohol [diol] 2,3 butanediol into MEK through a dehydration based reaction.[69] 2,3 butanediol can be derived from microbial based fermentation.[70] The combination of a natural organic acid such as butyric acid with ethanol forms the ester ethyl butyrate [3] while the mixture of levulinic acid with ethanol makes another ester ethyl levulinate [4].[66,71] Other examples include the natural fertilizer urea mixed with methanol to produce dimethyl carbonate [5][72] while glycerol mixed with acetic acid or isobutanol (isobutylene) forms the ether compounds of triacetal glycerol (TAG)[73] [6] & tritertiary butyl glycerol (TTBG) [7].[74] All of these compounds can ideally serve as oxygenated fuel additives that are derived from various types of industrial manufacturing schemes.

Synthesis of specific oxygenated fuel additives derived from reactions with natural compounds and/or alcohols

[1] isobutanol → isobutene + ethanol → ethyl tertiary butyl ether [ETBE]

[2] 2,3 butanediol → methyl ethyl ketone

[3] butyric acid + ethanol → ethyl butyrate

[4] urea + methanol → dimethyl carbonate (DMC)

[5] levulinic acid + ethanol → ethyl levulinate

[6] glycerol + acetic acid → triacetyl glycerol (TAG)

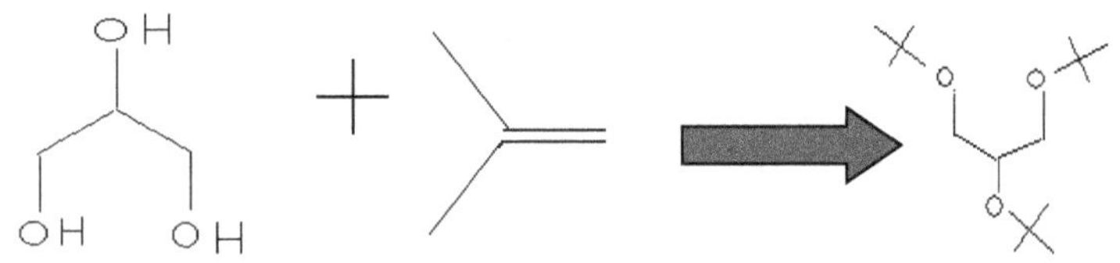

[7] glycerol   isobutene (from isobutanol)   tri-tertiary butyl glycerol (TTBG)

Figure 2-5 : The synthesis of oxygenated compounds made exclusively from alcohols include ethyl tert butyl ether (ETBE) [1] and methyl ethyl ketone (MEK) [2]. Just as important various esters and carbonates are made from the combination of a natural compound and an alcohol like ethyl butyrate [3], dimethyl carbonate [4] and butyl levulinate [5]. Other oxygenates include the combination of glycerol with acetic acid to form triacetyl glycerol (TAG) [6] and glycerol with isobutene (from isobutanol) (tri-tertiary butyl glycerol (TTBG) [7]).[66,68,69,71,72,73,74]

It should be known that some of these oxygenated compounds are more commonly utilized as fuel additives in diesel fuel formulations rather than for gasoline fuel. Some examples of oxygenates in diesel fuel include higher molecular weight ethers, esters, carbonates and glycerol based compounds. The addition of oxygenated compounds into diesel fuel have the potential to improve its cold temperature properties, lower pollutant types of emissions and enhance combustion performance by improving the cetane value and/or viscosity of the diesel fuel.

Table 2-5 demonstrates this point by showing the respective viscosities and melting points of certain oxygenated diesel fuel additives such as dibutyl ether, dimethyl carbonate, dimethoxymethane and ethyl acetate.[75,76,77,78] Each of these compounds make excellent diesel fuel additives as they have superior low melting temperatures that impart very good cold temperature behavior. In addition they contain very low viscosity values that decrease the incidences of unburned hydrocarbon accumulation within the engine, exhaust and lubrication system.

As stated previously, the alcohols are necessary ingredients for producing a large number of other oxygenated fuel compounds. Various alcohols serve as essential starting chemicals required for the manufacture of ethers, esters, carbonates and other organic fuel additive compounds. In addition, the alcohols themselves are effective as fuel additives. Examples include single higher molecular weight solutions like butanol, isobutanol or isopentanol. These alcohols can also be utilized directly as oxygenated fuel additives not requiring further chemical synthesis. Mixed higher molecular weight alcohols ranging in size distribution would also serve as a good gasoline fuel amendment. Mixed alcohol solutions can either consist of linear or branched alcohols. These alcohols are discussed further in chapter 6.

The viscosities and melting points of some common organic oxygenated compounds used as fuel additives

	Melting Point	Viscosity
Dibutyl ether	-95 °C	0.63 mm²/s
Dimethyl carbonate	-3 °C	0.63 mm²/s
Dimethoxymethane	-104 °C	0.37 mm²/s
Ethyl acetate	-83 °C	0.43 mm²/s

Table 2-5 : Several common oxygenated fuel additives that are ethers, esters or carbonates have very low viscosities and melting points that give the diesel fuel more enhanced performance properties.[75,76,77,78]

Ether compounds derived from renewables are another logical choice for fuel additives due to the large number of them that can be made from the alcohols. A dozen or more ether compounds can be produced from a combination of two different alcohols. The combinations of some alcohols that form ethers include methanol, ethanol, propanol, isopropanol, butanol or isobutanol. High molecular weight ethers are preferable to fashion into gasoline fuel additives. These ethers are less of an environmental health risk than those with a lower molecular weight such as Methyl Tertiary Butyl Ether (MTBE). Even though MTBE had health concerns due to environmental exposure, examination of the MSDS information on other types of ethers such as Tertiary Amyl Methyl Ether (TAME) & Ethyl Tertiary Butyl Ether (ETBE) denote that they have around 5 times less solubility in water than MTBE and are more easily removed from water and soil sources than MTBE.[79]

2.5 There are foreseeable problems associated with our current renewable fuel standards (RFS2)

The amount of biofuel produced within the United States is mandated or determined by laws. These laws are known as the Renewable Fuel Standards (RFS2) stipulated by the 2007 Energy and Independence Act. These standards set the allowable amounts of renewable fuels manufactured up to the year 2022. Figure 2-6 below summarizes the types of renewable fuels classified by RFS2, the allotted volume production of these fuel types by 2022 and the amount of carbon dioxide emissions permissible during their production.

Renewable fuel standards stipulate the volumes of biofuel manufactured and the relative amounts of carbon dioxide emitted

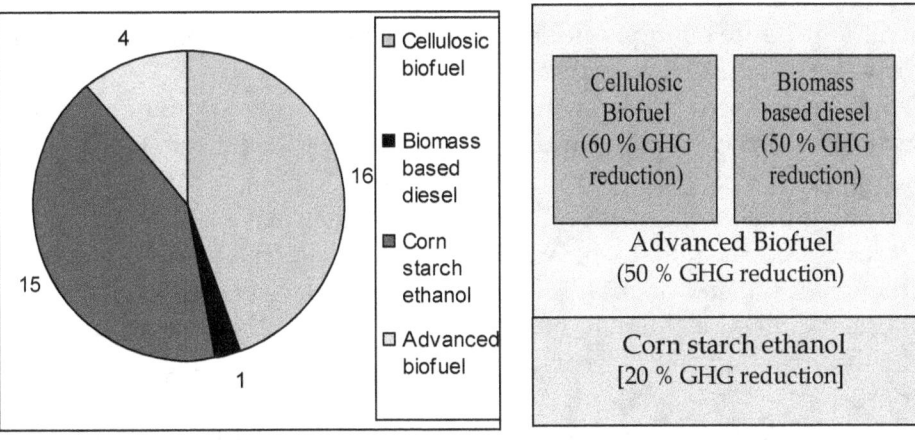

Figure 2-6 : The total volume of biofuel to be produced by 2022 was established as 36 billion gallons with the majority being either corn starch ethanol or renewable cellulosic biofuel.[80] Renewable biofuels must meet carbon dioxide emission standards that compare to those of petroleum fuel manufacture. Most renewable biofuel with the exception of corn starch ethanol must reduce carbon dioxide emissions by at least 50 % of the amount compared to petroleum fuel.[81]

According to the mandate, renewable biofuels are classified as 1) corn starch ethanol [renewable fuels category], 2) cellulosic biofuel such as renewable diesel or ethanol based fuel, 3) biomass based diesel and 4) advanced biofuel. The RFS2 standard decrees that the amount of total biofuel manufactured by 2022 should be 36 billion gallons divided up among the four classifications mentioned above. The majority of this amount, around 31 billion gallons per year, has been allocated to be split among corn starch ethanol and cellulosic biofuel production. The other biofuels include biomass based diesel produced and maintained at 1 billion gallons per year while advanced biofuels make up another 4 billion gallons per year. The advanced biofuel category includes other cellulosic based fuels but does not include renewable based diesel from sources like algae.[81]

Renewable fuel production is regulated by the EPA through end of the year evaluations and an accounting system that involves a Renewable Identification Number (RIN) required by the EPA.[81] Any biorefinery that produces over 10,000 gallons per year of biofuel must go through this accounting system if it is selling renewable fuel commercially.[81]

The setting of minimal or maximum limits for renewable fuel categories like cellulosic biofuels and biomass based diesel by the EPA has many disadvantages. Prime examples include setting a production cap on biomass based diesel as well as setting too high of a projected cellulosic biofuel volume. A set limit on biomass based diesel is not prudent since the demand for biodiesel already surpasses 1 billion gallons per year. It has also been difficult to produce enough cellulosic based biofuel as the amount required on a yearly basis to fulfill the established volumes for it.

In addition, the EPA evaluation guidelines also stipulate that other renewable feedstocks falling under the categories of cellulosic biofuels must be accepted as a new fuel production pathway before it can be produced. Under the guideline regulation, feedstock sources such as camelina, grasses and arundo donax had to be withdrawn from the final drafts for biofuel production at the end of 2011.[80] This rule may explain in part why the actual cellulosic biofuel production happens to be much lower than what was originally expected each year according to the projected yearly RFS2 standards.

Several additional complications arise from stipulating a set amount of biofuel from renewable sources. Primarily, the projected 31 billion gallons of ethanol produced from cellulosic or corn starch sources seem more than sufficient. The 13-14 billion gallons of it currently being produced fulfill the 10% mix (E10) requirement. Amounts of ethanol greater than the 10% additive to gasoline pose potential incompatibility issues. Some question the practicality of implementing a 15% ethanol gasoline (E15) fuel mixture. Their concerns involve the corrosion and phase separation issues that could take place with the higher ethanol to gasoline mixture in vehicles.

A second complication involves maintaining the projected manufacture of 15 billion gallons of corn starch ethanol per year after 2015. Dry grind mill based ethanol production already requires over three quarters of corn cultivated on US farmland.[82] The over establishment of corn cultivation on US farmland for fuel purposes has caused artificial price rises affecting major food crop commodities. Therefore, it would seem more practical to support alternative biofuel production from renewables other than corn in order to prevent further adverse food supply and price related issues. Funding alternative biofuel manufacturing technologies would allow for the gradual phasing out of ethanol made from corn starch. Such an approach would be a viable solution to the food-fuel problem we are currently facing.

A third complication associated with mandated fuel production concerns the allotted amount of renewable diesel obtained from biomass sources. This was originally set at 1 billion gallons per year up through 2022. Setting a limit on biomass based renewable diesel fuel has several drawbacks. The primary reason being that biodiesel manufacture has already surpassed the one billion gallon per year mark during the last several years [after 2011]. A set cap limit disallows other competing biodiesel manufacturers access to this burgeoning market.[83] In addition, a larger amount of biodiesel happens to be necessary towards satisfying B5 (5 % biodiesel mixture in gasoline) diesel fuel demand as shown on signs at fuel dispensing stations across the country. B5 fuel has certain advantages as a biofuel-petrodiesel mixture.

Another adverse factor in limiting biobased renewable diesel production has to do with the increasing need to make renewable diesel from algae. The cultivation and processing of algae biomass may become a technology necessary towards producing large volume amounts of renewable diesel in the future. The current RFS2 guidelines disallow algae based diesel to be considered as part of the advanced biofuel category. Therefore, its production capacity, along with vegetable oil or animal fat based biodiesel, are all capped at too low a maximum amount.

Last of all, most biofuel production technologies are mainly suited to make diesel fuel. In many of these circumstances, multiple types of vehicle fuels along with diesel are manufactured at one time. For example, pyrolysis and thermochemical gasification technologies can manufacture both diesel and gasoline fuels with the majority of it being renewable diesel. In the case of thermochemical gasification, low and high temperature processes mainly produce hydrocarbons in the range of 11 to 22 carbons in length.

The carbon dioxide issue. Carbon dioxide emission from biofuel production is also addressed in the RFS2 fuel mandate. The process of biofuel manufacturing itself emits sizable amounts of carbon dioxide. The more common biofuel manufacturing methods such as fermentation and thermochemical gasification converts at least 50 % or more of the incoming carbon source into carbon dioxide.[84,85] In addition, combustion of producer gas generating electrical power for the refinery emits carbon dioxide.

The US USDA-DOE based idea of an integrated biorefinery

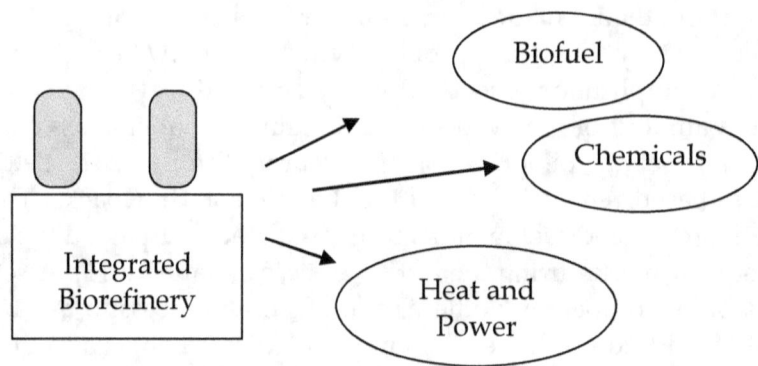

Figure 2-7 : An Integrated Biorefinery is defined as a refinery that mainly produces biofuel along with side products such as additional chemicals and also heat and power similar to what a petroleum refinery already does.

Integrated Biorefineries. The incorporation of future biofuel manufacture in integrated biorefineries includes plans to effectively capture carbon dioxide as well as increase energy efficiency. Appropriate carbon dioxide capture falls in line with the energy efficiency goals for integrated biorefinery designs. Government based funding projects such as the DOE Integrated Biorefinery Demonstration Project has supported the development of several alternative ethanol manufacture companies that are energy efficient and reduce carbon dioxide emissions. Therefore, continued funding of such refineries would be worth the time, money and effort due to the general progress of technological development associated with biofuel manufacture.

The production of ethanol or other biofuels (i.e., green gasoline or diesel) from fermentative, thermochemical or other alternative technologies fits the prescribed model or definition of what the DOE or USDA considers as an Integrated Biorefinery. The US government based definition of an Integrated Biorefinery is very similar to what a petroleum refinery does as is shown in Figure 2-7. An Integrated Biorefinery produces not only biofuel but important value added byproducts such as chemicals along with its own heat and power. The US government supports the development and construction of Integrated Biorefineries with guaranteed loans and grants through the USDA 2008 Farm Bill biorefinery assistance program[86] and the DOE demonstration of integrated biorefinery operations program under the 2009 American Recovery and Reinvestment Act.[87]

2.6 A large number of value added byproducts can be made from integrated biorefineries that utilize algae or saccharide sources

Plastics and chemicals derived from petroleum span an immense industry of everyday products that range from cosmetics to industrial paints and coatings. Using similar production methods, biobased chemicals from biorefineries have the potential to contribute just as much to these product based industries as the petroleum industry has done. These refineries would help to produce commodity chemicals that in turn help make essential ingredients in consumer and/or industrial products. The production of biobased chemicals not only assists a refinery in the recovery of plant operating costs but will allow monetary profits off such endeavors. This is already apparent with several companies that are manufacturing biofuel in conjunction with producing and marketing chemical byproducts.[88,89]

As a flourishing example, the soybean bio-industry has been responsible for producing a large number of consumer, food and industrial products from the soybean oil, meal and flour (Figure 2-8). Dozens of products such as paints, inks, textiles, pharmaceuticals, cosmetics, plastics and construction materials are manufactured from ingredients contained in the soybeans.[90] In addition, many everyday food products such as meat substitutes, snack foods, confections, cereals, diet foods and drinks commonly have soybean ingredients in them.[90] Well over one hundred products are made from ingredients taken from soybeans.

Similar to soybeans, algae cultivation & processing could potentially contribute towards the manufacture of a wide variety of similar products. Food products such as noodles, baby formula, ingredients in salad oil and drinks currently utilize bio-compounds derived from algae. Algae also give rise to a large number of neutraceutical & alternative health products as well as necessary ingredients in cosmetics. When combined with other compounds, the exopolysaccharides (sugars) emitted from algae have the potential to assist with the foaming, wetting, dispersion and solubilization processes of many manufactured products.[91] The contribution of ingredients from the exopolysaccharides have additional possibilities towards the manufacture of other synthesized products as is shown in Figure 2-9. The exopolysaccharides contain chemical components that assist in forming manufacturing ingredients such as emulsions, stabilizers, surfactants, foams, etc. These ingredients, in turn, could then produce finished synthesized products when combined with other ingredients. Examples of such finished products include coatings, plastics, paints, packaging, textiles, etc.

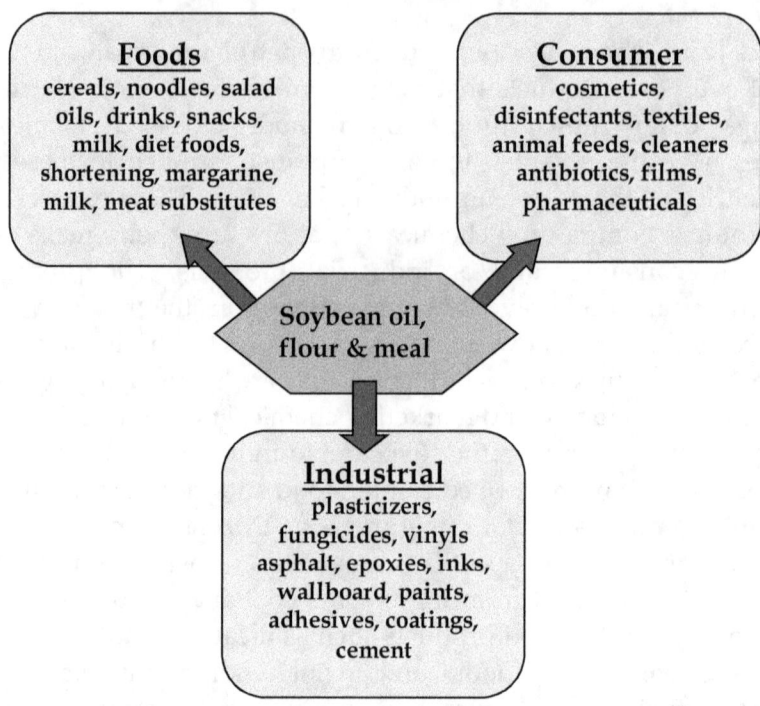

Figure 2-8 : A large number of food, consumer and industrial products are made from the oil, flour and meal of soybeans. Industrial products include plastics and construction materials while consumer products include inks, paints, textiles, cleaners, animal feeds, cosmetics, etc.[90]

In addition to a variety of cosmetics and health products, biocrude from algae provides opportunities for the production of other end use commodity chemicals. Due to the wide variety of algae species out there, an array of chemical building blocks can be produced from them. One example would be the extraction of fatty acids found in algae. These fatty acids are similar to those found in plant oilseed sources currently utilized for making products such as waxes, soaps, lubricants, cosmetics as well as medical, health and pharmaceutical products.

Biocompounds from algae can also be applied towards the manufacture of biodegradable plastics. Compounds such as lactic acid or polyhydroxyalkoanates are obtained through the cultivation and further processing of the algae under certain growth conditions.[92,93] However, in the case of lactic acid, algae would provide the saccharides necessary for its production through fermentation with other types of microbes.

Products made from the component ingredients obtained from the mixture of chemicals

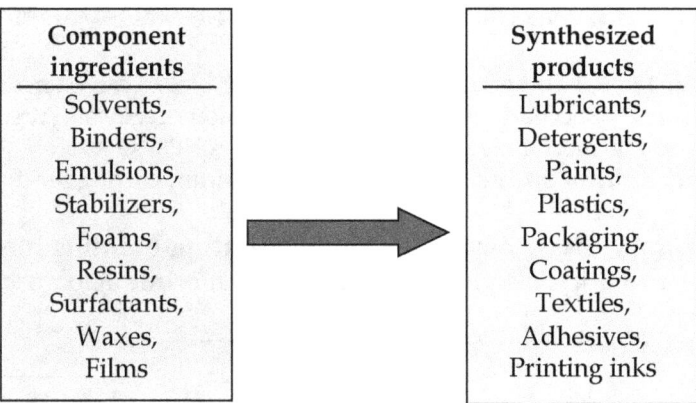

Figure 2-9 : There are necessary component ingredients required to make a variety of industrial products that include solvents, resins, waxes, surfactants, stabilizers, etc. These products are synthesized from certain chemicals mixed together with other components. Common synthesized industrial products include lubricants, detergents, textiles, inks, plastics, paints, etc.

As will be discussed in later chapters, biorefineries commonly produce biofuel from synthesis gas or the fermentation of saccharides from various sources. The Department of Energy in the past conducted a study nominating the top chemicals made from biorefineries that utilize gasification, fermentation or thermal conversion processes. The chemicals from these types of facilities could serve as the basic building blocks for the synthesis of common products.[94] Table 2-6 names some of these important chemical compounds that can be made from saccharides. Such biobased chemicals include succinic acid, aspartic acid, levulinic acid, sorbitol, itaconic acid, etc.[94] These building block chemicals are further processed into a larger array of other chemicals useful as ingredients required for the manufacture of consumer or industrial products.

As an example, Figure 2-10 demonstrates that the starting chemical building blocks of succinic acid and 3-hydroxypropionic acid each produce at least five or more compounds that can be input towards the manufacture of other useful products.[94] Through other chemical reactions, succinic acid can be converted to either g-butyrolactone, tetrahydrofuran, 2-pyrrollidone, 1,4 butanediol, etc.[95] The same chemical reaction schemes apply for 3-hydroxypropionic acid shown in the figure as well as the other chemicals mentioned in Table 2-6.

Important building block chemicals obtained from saccharides
Succinic acid, fumaric acid, malic acid, aspartic acid, glucaric acid, 3-hydroxypropionic acid, 2,5 furandicarboxylic acid, glutamic acid, itaconic acid, glycerol, sorbitol, levulinic acid, xylitol/arabinitol

Table 2-6 : Over 10 or so important chemical building blocks can be made originally from saccharides. Saccharides are common biomolecules encountered in fermentation or thermochemical based biorefineries. These basic chemicals can produce a larger array of other chemicals useful in manufacturing products.[94]

An array of chemicals made from the starting building blocks of 3-hydroxypropionic acid and succinic acid

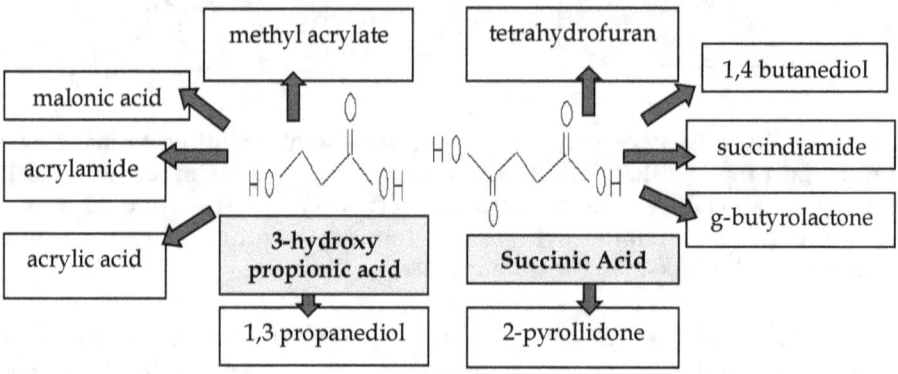

Figure 2-10 : 3-hydroxypropionic acid [left] and succinic acid [right] can create at least 5 or more other starting chemicals each used in further product synthesis.[94]

It is apparent that biorefineries have ample opportunities to manufacture a large array of marketable chemical byproducts beyond just the primary production of biofuel. For example, along with ethanol production, a thermochemical refinery can manufacture byproduct compounds such as isopropanol, acetone, hexane, acetaldehyde and butanol.[96] Hybrid ethanol refineries can produce compounds such as acetic acid, ethyl acetate, hexanol, propionic acid and lactic acid.[97] In addition, biorefineries that contain plasma enhanced melter [PEM®] technology can create vitreous solutions useful for recycled construction materials.[98] Also, bio-oil obtained from pyrolysis refineries can be further processed to extract the aromatic compounds not used in gasoline formulations. For example, the aromatic compounds of benzene, toluene and xylene (BTX) can produce an array of other solvents such as cyclohexane, ethyl benzene, acetone, benzaldehyde, phenol, benzyl alcohol and others.[99] All these essential chemical products will provide considerable value added revenue toward the successful operation of a biorefinery.

2.7 There are a large proportion of renewable feedstock sources useful towards biofuel production from algae cultivation, energy crop development and conversion of forestry wastes

A large amount of biofuel could be manufactured from the renewable resources that exist in the United States. These resources include the growth of energy crops, the recycling of crop residues, forestry residues, wood industry waste and algae cultivation. Algae growth and development could potentially provide enough biofuel for all our future vehicle fuel needs if the algae were given the appropriate resources required for its cultivation. This assertion was made by a researcher from the University of New Hampshire[100] who reviewed the findings of the Department of Energy's Aquatic Species Program conducted during several decades [1970's to 1990's] as shown in Figure 2-11.

According to the research program, it was estimated that a certain amount of biofuel called a 'quad' (equivalent to 7.5 billion gallons of fuel) could be produced within a land area consisting of 780 square miles when provided with enough carbon dioxide for algae cultivation.[101] It was then estimated that it would require a total of 19 quads of biofuel to satisfy total vehicle fuel requirements including diesel and gasoline fuels.[100] This corresponds to a total land area of approximately 15,000 square miles. To put this into perspective, this amount of land corresponds roughly to 13 % of the total area of the state of Arizona. In other words, cultivating algae on this amount of acreage would theoretically provide for all of our vehicle fuel needs given that the appropriate nutrient, water and carbon dioxide resources are provided for within this given area of cultivation.

Our existing land resources could also help produce biofuel when used appropriately in other ways. The US contains ample supplies of forestland, pastureland and farmland. The percentage distribution of these lands is 20 % for farmland, 28 % for forestlands and 26 % for pasturelands as is shown in Figure 2-12.[102] All three land ecosystems combined together account for approximately ¾ of all US land acreage. Pasturelands and forestlands have the potential to cultivate energy crops applicable towards biofuel production. In addition, these crops would not usurp valuable farmland resources. A simple calculation estimates that if 100 million acres of such land(s) were set aside for the cultivation of perennials, it could provide for 100 billion gallons of biofuel. This assertion assumes that 1000 gallons of fuel per year can be made from an acre of land cultivating the appropriate perennial crop. Such energy crops can also be grown on haylands or pastureland type of farmlands that may tend not to detract from the proper land cultivation required for producing normal food crops.

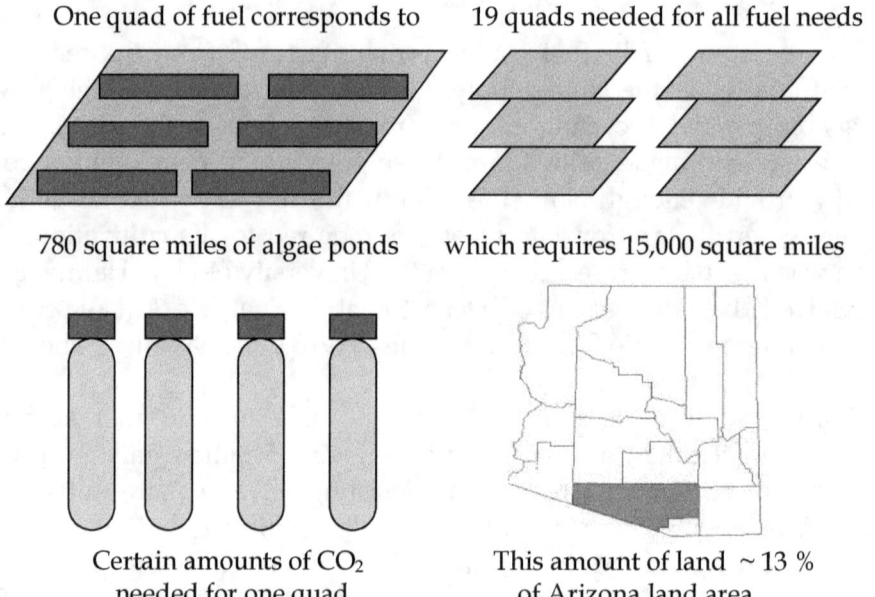

Figure 2-11 : The US DOE Aquatic Species Program estimated that large amounts of algae energy could be grown within a certain amount of land area called quads.[101] It was further shown by another researcher that around 19 of these quads or about 15,000 square miles, close to the size of Pima County, AZ, could satisfy all vehicle fuel demands.[100]

In addition to energy (perennial) crops, a large amount of biofuel can potentially be made from renewable wastes. Using waste resources in conjunction with growing energy crops would lessen the need to exclusively grow a large amount of energy crops for biofuel production. According to the US DOE & USDA billion ton study, just less than 1 billion tons of renewable wastes can potentially be processed each year for either energy, vehicle fuel or alternative products.[103] As shown in Figure 2-12, this amount is expected to come from either forestry waste residues (360 million tons per year), crop residues such as corn stover (250 million tons per year) and other renewable wastes (320 million tons per year).[103] In addition, the conversion of crop waste residues into biofuel does not potentially compete for food related purposes. The production of biofuel from corn stover is such an example. It's important to understand that a similar amount of ethanol or other biofuel could be made from the corn stover instead of from the starch contained in the corn.

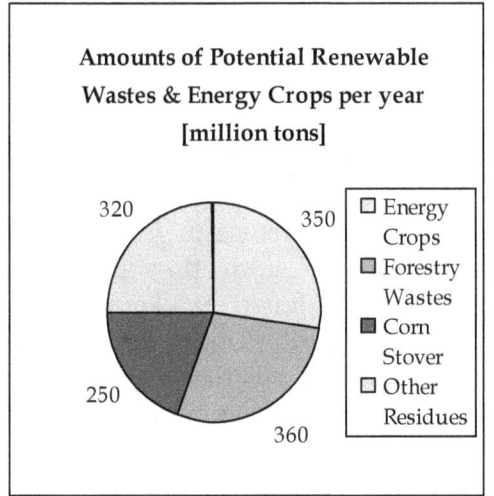
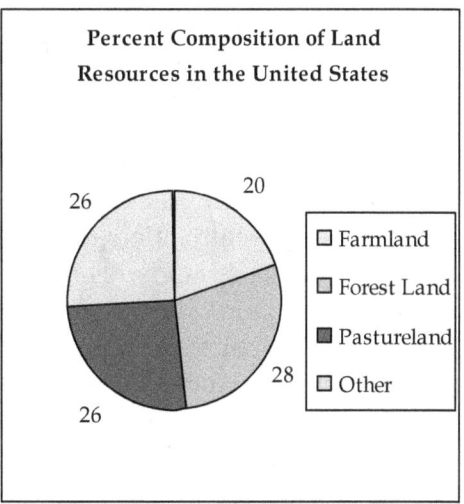

Figure 2-12 : A large amount of renewable wastes and resources potentially can be obtained from utilizing the wastes from forestry lands or rehabilitated lands of pastureland and marginal or abandoned farmland. Forest land and pastureland make up over 50 % (54 %) of the total US land acreage.[102] In addition, both forestry wastes and energy crops would contribute to over half of the renewable resources available each year according to the US DOE-USDA billion ton study.[103]

Approximately 360 million tons per year of forestry based renewable wastes are available for energy or fuel production. These forestry wastes include logging residues, land thinning or clearing, urban waste wood, removal of wood from forest fires, wood for electrical energy generation and unmerchandisable wood.[104] In addition, other types of forestry wastes can be obtained from lumber, paper/pulp and other wood processing mills as there are thousands of these businesses or outfits scattered across the country. These mills oftentimes process lumber into wood members, housing pieces [floors, doors, windows], specialize in sawing wood, make plywood, trusses, veneer and other reconstructed or specialty wood products,[105] discarding huge amounts of wooden scraps.

At least three types of convertible waste taken from forestry related mills can be converted into vehicle fuels. These include black liquor/tall oil, hydrolysate waste and various types of wood waste. Waste such as wood shavings, sawdust, bark and wood pieces from mills can be processed into biofuels through thermochemical gasification or pyrolysis types of refineries. Black liquor obtained from mill processing can be turned into energy or biofuel. In addition, liquors can then be further converted into tall oil, which can make biodiesel or other renewable fuels. Lumber mills also emit liquid waste known as hydrolysate from acid treatments. The hydrolysate can then be directly fermented into fuel alcohols such as ethanol or butanol. This type of processing removes the need to put lignocellulosic material through a pretreatment step since it is already done at the mill. However, the hydrolysate may still require further processing before being applied towards alcoholic fermentation.

Some potential circumstances exist where energy crop cultivation has sensible land applications as well as involvement in the production of biofuel. For example, the cultivation of energy crops can enhance or improve the condition of certain lands. This can take place by cultivating grasses or perennials for land remediation purposes. As shown in Figure 2-13, remediation of pasturelands/farmlands include restoring lands that have soil erosion problems, wetlands that have been drained, lands that require water quality improvement by preventing circumstances such as nutrient runoff, maintaining wildlife habitat by using cover crops and preventing overgrazing by cultivating grasses.[106,107]

Grasses and perennials can be grown to remediate lands that require soil and water quality improvement and prevent overgrazing

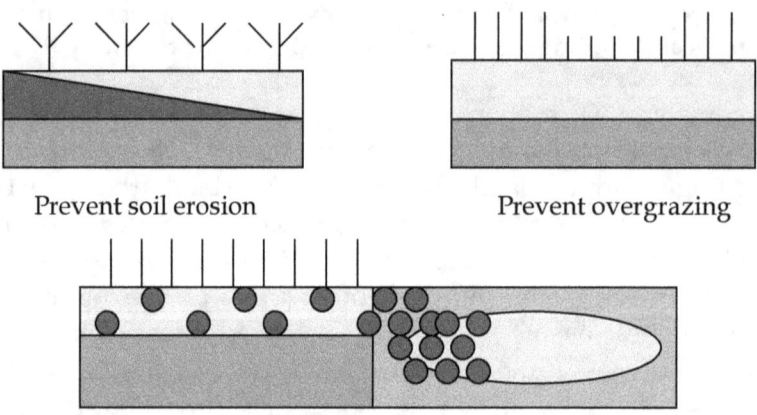

Prevent soil erosion   Prevent overgrazing

Prevent nutrient runoff into water bodies

Figure 2-13 : There are many circumstances where farmlands and pasturelands require remediation such as improving soil conditions due to erosion, preventing animal overgrazing and improving water quality by preventing nutrient runoff into water bodies.[106,107]

One government program called the Conservation Reserve Program [CRP], has already set aside 30 plus million acres of farmland and pastureland in need of remediation.[107] Such programs allow landholders to lease the land for a period of 10 -15 years in order to improve the condition of their existing land resources.[106] These landowners oftentimes cultivate grasses, trees or forage plants for the aforementioned land restoration purposes.

Cultivation of specific wood energy crops may also be of interest especially concerning those types that grow rapidly. These crops could also apply towards land rehabilitation or other environmental applications. Such examples include hybrid poplars, aspen and birch trees. Hybrid poplars, created from crosses of certain cottonwoods or poplars, grow very rapidly. Hybrid poplar wood chips have been used in several biofuel production models in the past.[108,109] Another set of potential energy crops include aspen and birch trees that are grown on rehabilitated lands that oftentimes include regenerating forestland claimed by fires. These trees also grow rapidly and are ideal for growth in areas without current vegetation cover.

REFERENCES:

42. Status and impact of state MTBE Ban, http://www.eia.gov/oiaf/servicerpt/mtbeban - accessed May 2014
43. **Motor Gasolines Technical Review** – by L. Gibbs, B. Anderson, K. Barnes et al [2009] Chevron Corporation, p.6, 47-48
44. **Gasoline – What's that stuff?** – Chemical and Engineering News Vol 83 No 8 pg 37 [2005] by S. Ritter
45. **Structural group contribution method for predicting the octane number of pure hydrocarbon liquids** – Industrial and Engineering Chemistry Research Vol 42 pgs 657 – 662 [2003] by TA Albahri
46. **High octane components from catalytic reformates** – Industrial and Engineering Chemistry Vol 51 No 1 pgs 73 – 76 [1959] by AD Reichle
47. **The Shell middle distillate synthesis process (SMDS)** – Catalysis Letters Vol 7 no 1 – 4 pgs 253 – 269 [1991] by J. Eilers, SA Posthume, ST Sie
48. **Optimal process conditions for the isomerization – cracking of long chain n-paraffins to high octane isomerizate gasoline over Pt/SO4-ZrO2 catalysts** – Fuel Processing Technology Vol 92 pgs 1675 – 1684 [2011] by M. Busto, CR Vera, JM Grau
49. Process description and performance of ExSact – http://www.exelusinc.com/exsactperformance.shtml - accessed May 2014
50. **Compression ignited homogeneous charge compression** – SAE Technical Paper 830264 by D. Najt, D. Foster [1983]
51. **Combustion control method of homogeneous charge diesel engines** – SAE Technical Paper 980509 by H. Suzuki, N. Koika, M. Odaka [1998]

52. **Development of a gasoline engine system using HCCI technology – the concept and test results** – SAE Technical Paper 2002-01-2832 by J. Yang, T. Culp, T. Kenney [2002]
53. **HCCI combustion phasing in a multiple cylinder engine using variable compression ratio** – SAE Technical Paper 2002-01-2858 by G. Haraldsson, P. Tunestal, P. Johansson, J. Hyvonen [2002]
54. **Experimental study of combustion and emission characteristics of ethanol fueled port injected homogeneous charge compression ignition (HCCI) combustion engine** – Applied Energy Vol 88 No 4 pgs 1160 – 1180 by RK Maurya, AK Agarwal
55. **Demonstrating the multi-fuel capability of a homogeneous charge compression ignition engine with variable compression ratio** – SAE Technical Paper 1999-01-3679 [1999] by M. Christensen, A. Hultqvist, B. Johansson
56. **An investigation of hydrogen fuelled HCCI engine performance and operation** – International Journal of Hydrogen Energy Vol 33 pgs 5823 – 5828 [2008] by JMG Antenes, R. Mikelson, AP Roskilly
57. **Numerical study of the combustion mechanism of a HCCI engine fueled with dimethyl ether and methane, with a detailed kinetics model : Part 1 – The reaction kinetics of dimethyl ether** – Proceeding of the Institute of Mechanical Engineers : Part D Journal of Automotive Engineering Vol 219 pgs 1213 – 1223 [2005] by
58. **Experimental study on HCCI combustion of dimethyl ether (DME) / Methanol dual fuel** – SAE Technical Paper 2004-01-2993 [2004] by ZQ Zhang, MI Yao, Z Chen, B Zhang
59. **Comparison of simulated and experimental combustion of biodiesel blends in a single cylinder diesel HCCI engine** – SAE Technical Paper 2007-01-4010 by JP Szybist, J McFarlane, BG Bunting
60. **Combustion characteristics of tricomponent fuel blends of ethyl acetate, ethyl propionate and ethyl butyrate in homogeneous charge compression ignition (HCCI)** – by F. Contino, F. Foucher, C Mouaim-Rouselle
61. **Use of a single-zone thermodynamic model with detailed chemistry to study a natural gas fueled homogeneous charge compression ignition engine** – Energy Conversion and Management Vol 53 No 1 pgs 298 – 304 [2012] by J. Zhang, JA Caton
62. **Increase of the environmental and operational characteristics of automobile gasolines with the introduction of oxygenates** – Theoretical Foundations of Chemical Engineering Vol 43 No 4 pgs 563-567 [2009] by AV Tsarev, SA Karpov
63. **Evaluation of butanol-gasoline blends on a portable fuel injection spark ignition engine** – Oil & Gas Science and Technology Vol 65 No 2 pgs 345-351 [2010] by J. Dernotte, C. Mounaim-Rouselle, F. Halter, D. Seers

64. **Evaluation of the impact of fuel hydrocarbons and oxygenates on groundwater resources** – Environmental Science and Technology Vol 38 No 1 pgs 42 – 48 [2004] by T. Shih, Y. Rong, T. Harmon, M. Suffet

65. **Isobutanol – A renewable solution for the transportation fuels value chain** – GEVO White Paper [2011] http://gevo.com/assets/pdfs/GEVO-wp-iso-ftf.pdf - accessed May 2014

66. **An examination of the biorefining process, catalysts and challenges** – Catalysis Today Vol 145 pgs 138 – 145 [2009] by DJ Hayes

67. **Study of the combustion and emission characteristics of a diesel engine operated with dimethyl carbonate** – Energy and Conversion Management Vol 47 No 11 – 12 pgs 1438 – 1448 [2006] by L. Xiaolu, L. Hongyan, Z. Zhiyong, H. Zhen

68. **Kinetics of the liquid phase synthesis of Ethyl ter-Butyl Ether (ETBE)** – Industrial and Engineering Chemistry Research Vol 33 pgs 581-591 [1994] by C. Fite, M.Iborra, J. Tejero et al

69. **The dehydration of fermentative 2,3 butanediol into methyl ethyl ketone** – Biotechnology and Bioengineering Vol 29 pgs 343-351 [1987] by NV Tran, RP Chambers

70. **Microbial 2,3 butanediol production : A state of the art review** – Biotechnology Advances Vol 29 No 3 pgs 351-364 [2011] by H. Huang, PK Ouyang

71. **The future of butyric acid in Industry** – The Scientific World Journal Vol 2012 article ID 471417 10 pgs M. Dwidar, JY Park, RJ Mitchell, BI Sang

72. **Synthesis of dimethyl carbonate from urea and methanol over ZnO(x)-CeO2(1-x) catalysts prepared by sol-gel method** – Journal of Industrial and Engineering Chemistry Vol 18 No 3 pgs 1018 – 1022 [2012] by W. Joe, HJ Lee, UG Hong, YS Ahn et al

73. **Acidic mesoporous silica for the acetylation of glycerol : synthesis of bioadditives fo petro fuel** – Energy and Fuels Vol 21 pgs 1782 – 1791 [2007] by JA Melero, R. van Gricken, G. Morales, M. Paniagua

74. **Design and control of the glycerol tertiary butyl ethers process for the utilization of a renewable resource** – Industrial and Engineering Chemistry Research Vol 50 No 22 pgs 12706-12716 [2011] by JK Chang, CL Lee, YT Jhuang, JL Ward, IL Chien

75. **Methylal (melting point & viscosity) from properties section** – http://www.microkat.gr/msdspd90-99/Methylal.htm -- accessed May 2014

76. **ethyl acetate (melting point & viscosity) from section 9 - physical and chemical properties of MSDS sheet** – http://babec.org/files/MSDS/EtAc.pdf -- accessed May 2014

77. **Tables for Chemistry – liquid properties of dibutyl ether** – http://www.stenutz.eu/chem/solv6.php?name=dibutyl%20ether – accessed May 2014

78. dimethyl carbonate melting point and viscosity from section 9 (physical and chemical properties) of MSDS sheet - http://www.dimethylcarbonate.com/dimethyl_carbonate_msds.html -- accessed May 2014

79. MSDS sheets for tert-amyl methyl ether (TAME), ethyl tert-butyl ether (ETBE) and methyl tert-butyl ether (ETBE) from Fischer Scientific - online MSDS pages

80. Renewable Fuel Standards Program (RFS) - US EPA Office of Transportation and Air Quality [May 2012] by R. Argyropoulos

81. Renewable Fuel Standards (RFS) : Overview and issues - Congressional Research Service [Mar 2013] by R. Schnepf, BD Yacobucci - http://www.fas.org/sgp/crs/misc/R40155.pdf -- accessed May 2014

82. US World of Corn 2011 - by the National Corn Growers Association

83. EPA proposes 2012 Renewable Fuel Standards and 2013 biomass based diesel volume - http://www.epa.gov/otaq/fuels/renewablefuels/420f11018.pdf - accessed May 2014

84. Biofuels and Bioenergy : Process and Technology [Green Chemistry and Chemical Engineering] - p.71 (Ethanol from Corn) [2012 CRC Press] by S. Lee & YT Shah

85. Pressure swing adsorption for CO2 capture in Fischer-Tropsch fuels production from biomass - Journal of the International Adsorption Society Vol 17 No 3 SI pgs 443-452 [2011] by AM Ribeiro, JC Santos, AE Rodrigues

86. 2008 Farm Bill side by side - title ix - biorefinery assistance - http://www.ers.usda.gov/farmbill/.. -

87. Recovery Act - Demonstration of Integrated biorefinery operations - http://www07.grants.gov/...

88. Market areas/chemicals - creating renewable oils for the chemical market - http://solazyme.com/chemicals - accessed May 2014 (no longer available)

89. Polymer and plastic additives - http://www.amyris.com/products/163/PolymersPlastics - accessed May 2014 (no longer available)

90. Soybean uses - What can be made out of soybeans - Iowa State University Soybean Extension and Research Program - Mar 2007 - http://extension.agron.iastate.edu/soybean/uses_soyproducts.html - accessed May 2014

91. Biosurfactants, bioemulsifiers and exopolysaccharides from marine microorganisms - Biotechnology Advances Vol 28 No 4 pgs 436-450 [20xx] by SK Satpute, IM Banat, PK Dhakephalkan, AG Banparkar et al

92. **Investigation of utilization of the algal biomass residue after oil extraction to lower the production cost of biodiesel** – Journal of Bioscience and Bioengineering Vol 114 No 3 pgs 330-333 [2012] by MT Gao, T. Shimamura, N. Ishida, H. Takahashi

93. **Biosynthesis and mobilization of poly(3-hydroxybutyrate) [P(3HB)] by spirulina platensis** – International Journal of Biological Macromolecules Vol 36 pgs 144 – 151 [2005] by MH Jau, SP Yew, PSY Toh, ASC Chong et al

94. **Top value added chemicals from biomass : Vol 1 – Results of screening for potential candidates from sugars and synthesis gas** – NREL & PNNL [20xx] by T. Werpy, G. Petersen, A. Aden, J. Bozell et al

95. **The Integrated bio-refinery : Conversion of corn fiber to value added chemicals** – 4rth World Congress on Industrial Biotechnology & Bioprocessing [Mar 2007] by S. Kleff and MBI International

96. **Heterogeneous catalytic conversion of dry syngas to ethanol and higher alcohols on Cu based catalysts** – ACS Catalysis Vol 1 No 6 pgs 641 – 656 [2011] by M. Gupta, ML Smith, JJ Spivey

97. **Zeachem Technology – Products section** – http://www.zeachem.com/technology/products.php - accessed May 2014 (no longer available)

98. **InenTec | PEM® Technology --> Process Details** – http://www.inentec.com/pemtm-technology/process-details.html - accessed May 2014

99. **Life cycle inventories of petrochemical solvents and highly pure chemicals** – Swiss Center for Life Cycle Inventories – 2nd International ecoinvent meeting [Mar 2008] by G. Wernet, J. Sutler

100. **Widescale biodiesel production from algae** – University of New Hampshire Physics Dept – UNH Biodiesel group [2004] by M. Briggs

101. **A look back at the US Department of Energy's Aquatic Species Program – Biodiesel from algae** – National Renewable Energy Laboratory (NREL) NREL/TP-580-24190 [July 1998] by J. Sheehan, T. Dunahay, J. Benemann, P. Roessler

102. **Major uses of land in the United States, 2002** – USDA Economic Research Center – Economic Information Bulletin No EIB-14 [2006] by RN Labowski, M. Versterby, S. Bucholtz et al

103. **Biomass as a feedstock for a bioenergy and bioproducts industry : The technical feasibility of a billion ton annual supply – Factors increasing biomass resources from agriculture** – by US Department of Energy & US Department of Agriculture – R. Perlack, L. Wright, A. Turhollow, R.Graham, B.Stokes et al [Apr 2005]

104. **An update to the billion ton study annual supply** – Department of Energy [Sept 2010] by B. Perlack, B. Stokes, C. Eaton et al

105. 1997 Economic Census – [Manufacturing-Industry Series] – Engineered Wood Members, Truss Manufacturing, Reconstituted Wood Product Manufacturing, Wood Window & Door Manufacturing, Other Millwork – http://www.census.gov/epcd/www/ind31.htm – accessed May 2014 (no longer available)

106. Conservation Reserve Program Fact Sheet – USDA Farm Service Agency [Feb 2013] – http://www.fsa.usda.gov/Internet/FSA_File/crpfactsheet0213.pdf -- accessed May 2014

107. Conversation Reserve Program : Status and current issues – Congressional Research Service [Sept 2010] by T. Cowan

108. A model of wood flash pyrolysis in fluidized bed reactor – Renewable Energy Vol 30 pgs 377-392 [2005] by Z. Luo, S. Wang, K. Cen

109. The production and evaluation of oils from the steam pyrolysis of poplar chips – 8 pgs by DGB Boocock, A Chowdhury, SG Allen

Chapter 3

Renewable Fuels from Thermochemical Processing

3.1 Thermochemical production technologies effectively utilize renewable resources towards the manufacture of biofuels

Those interested in learning the basics of biofuel production will undoubtedly want to become familiar with thermochemical based fuel technologies. Knowledge of thermochemical methods forms a sound basis for the further study of biofuel production. For example, current hybrid biofuel production technologies involve some type of thermochemical based principal of operation. These operations are oftentimes mixed with other production methods such as fermentation.

The importance of thermochemical biofuel production from biomass based feedstocks cannot be over stated. It currently leads the development towards increasing biofuel manufacturing capacity in general since it happens to be one of the few available technologies capable of producing biofuel on a large volume scale. Since this book concentrates on presenting alternatives to ethanol corn starch manufacture, the manufacture of biofuel from thermochemical methods is definitely a viable solution to this problem. The principle biofuels that have the potential to be produced from thermochemical processing on an industrial scale are renewable diesel and alcohols like ethanol.

Thermochemical based technologies include thermochemical gasification, pyrolysis and liquefaction. During World War II the Germans developed the Fischer-Tropsch method of producing vehicle fuel from coal resources. Even today, the principal thermochemical gasification technology utilized still tends to be Fischer-Tropsch synthesis. This technology operates on the principles of generating a synthesis gas from biomass or fossil fuels and then converting it into liquid fuels with the aid of metal catalysts.

Pyrolysis is also a thermochemical method centered upon producing volatile organic compounds that later become converted into bio-oil upon subsequent condensation. Synthesis gas and bio-oil are the two main intermediate products made from thermochemical technologies. After further refinement these intermediates then become the final biofuel product. Thermochemical gasification and pyrolysis are also very flexible in application. They are able to produce several vehicle fuel types at one time as well as implement a wide range of feedstock sources towards biofuel production.

Pyrolysis and liquefaction are similar thermochemical technologies with the difference being that liquefaction takes place in the presence of a liquid medium. Currently pyrolysis is more commonly practiced than liquefaction. However, this situation may change in the future since liquefaction preferentially processes high moisture content feedstocks such as algae.

All of the renewable wastes and energy crops mentioned in the USDA/DOE billion ton study can be applied towards biofuels production from thermochemical technologies. The potential for utilizing a wide array of biomass based feedstocks during processing make thermochemical gasification, pyrolysis and liquefaction very attractive biofuel manufacturing methods. In this regard, it becomes beneficial for a refinery to accept a multitude of feedstock sources since this reduces the need to locate and transport a certain specific biomass feedstock from a distant location. This practice should also reduce the upfront purchasing costs of the feedstock materials themselves.

Additional advantages for choosing thermochemical based technologies for biofuel production include recycling generated wastes emitted by the refinery and allowing a refinery to produce its own electrical power and process heat needs. Recycled generated wastes from refineries usually consist of char, effluent gases and carbon dioxide. The ability of refinery to generate its own power and heat saves on operating costs and lessens our reliance on fossil fuels. This concept of sustainability also fits in well with the integrated biorefinery model discussed throughout the book.

3.2 Thermochemical gasification can produce many types of vehicle fuels and is well suited for renewable diesel production

A wide array of vehicle fuels from biomass or alternative fossil fuel sources, like natural gas or coal, can be produced through thermochemical gasification. The main type of thermochemical gasification production technology is Fischer-Tropsch synthesis (FTS). Fischer-Tropsch synthesis can be classified as biomass to liquids (BTL), gas (i.e., methane) to liquids (GTL) or coal to liquids (CTL) depending upon the type of feedstock utilized. This publication mainly centers upon biomass to liquids (BTL) Fischer-Tropsch synthesis.

Fischer-Tropsch synthesis (FTS) manufactures a wide variety of compounds that can be classified into several types of vehicle fuels as is displayed in Figure 3-1. Vehicle fuels made from the FTS process include gasoline, jet fuel/diesel range hydrocarbons, oxygenates and gases similar to liquefied petroleum gas (LPG)[110]. For the diesel/jet fuel cut, there is a choice during the refinement process of producing paraffins for diesel fuel or kerosene compounds for jet fuel.[111] Gasoline type of hydrocarbons ranging in size from five to ten (C_5 to C_{10}) carbons are a straight cut that can be isolated or removed from the overall process. This cut is usually called naphtha. Hydrocarbons similar to Liquefied petroleum gas (LPG) are another biofuel cut made from FTS. It consists of volatile hydrocarbons that range from methane (C_1) to butane (C_4), but can also consist of alkene (double bonded) volatile compounds. In addition, a set of mixed oxygenates that contain compounds like alcohols, ethers, etc are oftentimes produced as well. Thus, the FTS process can make a wide array of compounds simultaneously or manufacture a specialized set of compounds if desired.

The various vehicle fuel cuts that can be obtained from the Fischer-Tropsch synthesis process

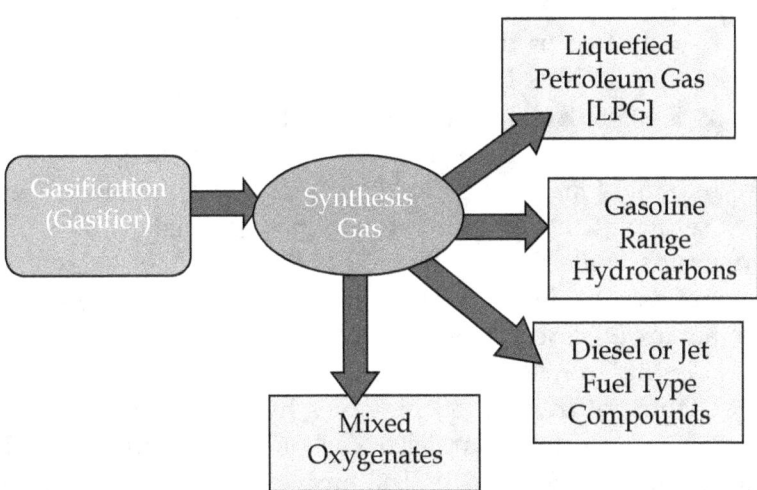

Figure 3-1: Biomass can be converted into almost all types of vehicle fuels manufactured at one time or separately using a Fischer-Tropsch process.[110]

Most importantly, all of these vehicle fuels from FTS have the potential to be made from a wide variety of biomass feedstock sources. Therefore, it is encouraging that thermochemical gasification itself has the flexibility of converting several different types of biomass or municipal waste into vehicle fuels. This includes lignocellulosic based wastes such as crop residues (corn stover), forestry wastes, organic (food wastes), energy crops like switchgrass and also municipal waste sources that include plastics and other cellulosic based commodities (paper & cardboard).

Themochemical gasification employs the use of a gasifier in order to make synthesis gas components (a combination of hydrogen and carbon monoxide). The gasifier thermally heats up and breaks down biomass into a range of various gases and other impurities such as tars and particulates. The gases produced are not only synthesis gas components but other types as well. However, the other gases must either be removed or become converted back into synthesis gas. The cleanliness of the synthesis gas relies upon removing the tar and particulate impurities.

Once the above mentioned processes are executed the gaseous product stream should consist of just hydrogen and carbon monoxide. The final consistency or proportion of these two synthesis gas components one to another, can be varied during its production. The final synthesis gas consistency primarily determines what type(s) of vehicle fuels are being made.

After their production, the synthesis gas components collect onto the surfaces of metal catalysts. There, subsequent chemical reactions take place, resulting in the formation of hydrocarbons that eventually make up the vehicle fuel. Fischer-Tropsch catalysts consist of several metals combined together. Catalyst materials are composed of metals like iron, nickel, cobalt and copper. Usually two or three of these base metals make up the primary components of the catalyst. These basic metals are then mixed with smaller amounts (i.e., < 5 %) of alkali or transition metals.

In general, Fischer-Tropsch synthesis can be divided into low and high temperature refining. One process method is normally favored over another in a refinery setting. The high temperature process makes lower molecular weight compounds while the low temperature one produces high molecular weight compounds. The low temperature method produces waxy paraffins compounds that are 20 carbons or more in length. This process is executed at temperatures ranging from 200 – 240 ° C.[112]

The high temperature method produces hydrocarbons with size ranges of several carbons up to around 20 carbons in length, carried out at temperatures from 300 – 350 ° C. [112] The high temperature process is usually preferred since it makes hydrocarbons suitable for most vehicle fuel types. The low temperature process makes hydrocarbons too high in molecular weight for direct use as a vehicle fuel. Therefore, these compounds must go through a subsequent hydrocracking process in order to break them down into smaller ones.

It is interesting to know that diesel fuel can be manufactured from FTS by combining the hydrocarbon fractions made from the low and high temperature production processes. This takes place by mixing a straight run fraction from the high temperature process with a refined portion originating from the low temperature method as is shown in Figure 3-2. The high temperature process usually gives off two fractions which are naphtha and a straight run diesel fraction. The straight run portion requires no further refining and can be used directly as diesel fuel. The naphtha fraction usually gets removed and becomes further processed until it gets converted into gasoline fuel. On the other hand the low temperature process produces an array of high molecular weight hydrocarbons averaging over 20 carbons in length. After initial production this fraction requires further refinement before it can be combined with the straight run portion in order to make usable diesel fuel. This post processing step takes place in a separate reactor that executes hydrocracking and hydrotreatment processes within it.

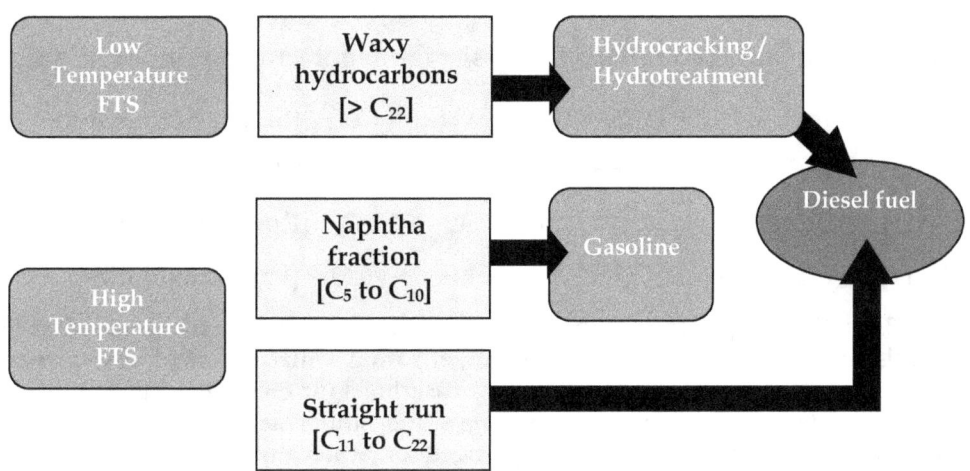

Figure 3-2 : A diesel fuel can be made by combining the separate fractions made through the low and high temperature processes of Fischer-Tropsch synthesis.[119] Some of these cuts require further post-processing. Hydrocracking breaks down larger waxy hydrocarbons from the low temperature FTS process into smaller ones.

3.3 The appropriate amounts of synthesis gas components are determined by specific chemical reactions and gasifier reactor conditions

Carbon, oxygen, and hydrogen are the essential elemental components of biomass. These components are transformed into synthesis gas through a set of chemical reactions that involve the combination of the biomass with water and oxygen done at high temperatures. Although there are many types of gas generation reactions, the four basic types shown in Figure 3-3 are responsible for the creation of a majority of the synthesis gas. These four reactions are **steam reforming, water shift, reverse water shift** and **partial oxidation**.

The partial oxidation reaction directly produces hydrogen and carbon monoxide by having hydrocarbons react with oxygen. Steam reforming produces both hydrogen and carbon monoxide upon the reaction of hydrocarbon based biomass components with steam (i.e., water). Elemental carbon can also react with steam in order to form these synthesis gas components. The water shift & reverse water shift reactions convert carbon dioxide or carbon monoxide into one of the synthesis gas components. The water shift reaction forms hydrogen and carbon dioxide upon reaction of some carbon monoxide with water. On the other hand the reverse water shift reaction makes carbon monoxide and water from the combination of carbon dioxide and hydrogen. In summary, steam reforming and partial oxidation produce the basic synthesis gas components from the biomass itself while the water shift reactions help to control the amounts of synthesis gas components upon reaction with other gases.

The basic chemical equations that assist in manufacturing synthesis gas

[1] C_x [or C_nH_m] + $H_2O \longrightarrow xH_2 + yCO$ (Steam Reforming)

[2] $CO + H_2O \longrightarrow CO_2 + H_2$ (Water Shift)

[3] $CO_2 + H_2 \longrightarrow CO + H_2O$ (Reverse Water Shift)

[4] $C_nH_m + O_2 \longrightarrow xH_2 + yCO$ (Partial oxidation)

Figure 3-3 : Four types of reactions are primarily involved in the generation of synthesis gas (carbon monoxide and hydrogen) from a mixture of biomass, steam and oxygen. Steam reforming converts hydrocarbons (or elemental carbon) mixed with water directly into synthesis gas. The water shift reactions combine carbon dioxide or carbon monoxide with hydrogen or water to produce one of the synthesis gas components. Partial oxidation combines hydrocarbons with oxygen to also make synthesis gas.

In addition to these reactions, effective synthesis gas generation also involves the right reaction conditions of temperature, steam and air or oxygen. The operational parameters of temperature, gas flow rate and steam to biomass ratio as shown in Table 3-1 help to determine the efficiency of producing optimal amounts of synthesis gas components. They also control the amount of impurities produced in the gas stream.[113] As shown in the table, these parameters affect the amounts of hydrogen and carbon monoxide generated. In addition, when they are applied in the correct manner, they reduce the amounts of char, soot, tars and carbon dioxide emitted.

Another point concerning synthesis gas generation is that a refinery oftentimes will want to control the ratio of the synthesis gas components produced (i.e., the amount of hydrogen relative to the amount of carbon monoxide). Synthesis gas ratios are effectively controlled through combining the processes of [1] steam reforming and [4] partial oxidation in a reactor. Steam reforming provides higher amounts of hydrogen while partial oxidation yields more carbon monoxide gas.

The three gasifier operational parameters that
optimize gas synthesis and reduce impurities

	Results
Gas Flow Rate Temperature (600 – 900 ° C), Steam:Biomass Ratio	Affects amounts of Hydrogen, carbon monoxide; Reduces carbon dioxide, char, soot, tar

Table 3-1 : Operational parameters determining effective production of synthesis gas as well as determined amounts of gas impurities.[113]

A secondary reformer combines fresh synthesis gas with recycled effluent
gases to adjust the consistency of hydrogen and carbon monoxide

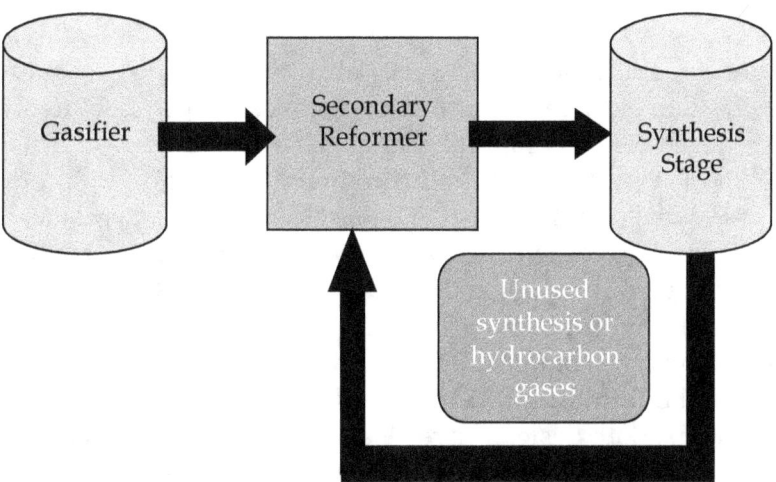

Figure 3-4 : The recycling of the waste effluent gases can adjust the consistency of synthesis gas components through a feedback loop that delivers the gases back into a secondary reactor. This reactor then combines the waste gases with the appropriate reactants (i.e., steam or oxygen) to form more carbon monoxide or more hydrogen according to the reactions [1-4] shown in Figure 3-3.

The synthesis gas ratio is oftentimes varied or optimized in a refinery between the values of 0.5:1 to 2:1 by controlling the amounts of steam, oxygen and/or volumetric flow rate of feedback waste gases. The low temperature Fischer-Tropsch process usually requires a high synthesis gas ratio (over 1:1) while the high temperature method normally gets provided with lower synthesis gas ratios (close to 1:1 or less). Therefore, the correct chemical constituency of synthesis gas is essential in helping to determine the types of products a refinery desires to make.

In addition, recycling control methods are oftentimes established in a refinery in order to help control synthesis gas ratios. One such method involves the recycling of waste effluent gases as a feedback loop process shown in Figure 3-4. The waste effluent gas components of carbon dioxide, methane, unused synthesis gas and hydrocarbons are recycled back into a **secondary reformer** in a feedback loop. In effect, this method adjusts the ratio of hydrogen or carbon monoxide by controlling the basic reactions [1-4] outlined earlier in this section. In essence this process combines the synthesis gas coming out of the gasifier with the recycled effluent gases. Steam oftentimes is added to the secondary reformer as well.

3.4 There are a variety of gasifier configurations utilized in thermochemical gasification

In order to appreciate how renewable biofuels are produced from thermal gasification one should learn the principles of operation involving the equipment utilized for making and cleaning the synthesis gas. This section covers the most essential piece of equipment necessary to produce the synthesis gas, which is the gasifier. The basic type of gasifier system implemented for thermochemical gasification is the **fluidized bed reactor**. This type of reactor configuration can also be outfitted for pyrolysis based biofuel generation covered in later sections.

Two common types of fluidized bed reactors utilized for thermochemical gasification are the **bubbling fluidized bed (BFB)** and **circulating fluidized bed (CFB)** reactors. Both designs have a heating media such as sand, plus they incorporate a fluidizing agent such as air, steam or oxygen-steam. Fluidizing agents can also consist of high pressure oxygen provided by an **Air Separation Unit (ASU)** or recycled effluent gases formed further down in the refinery. The fluidizing agents facilitate the chemical reactions necessary to make the synthesis gas components. The reactors are configured to run at atmospheric or higher pressures during operation.

The circulating fluidized bed reactor oftentimes has a **char combustor** connected to it. The operation of the circulating fluidized bed (CFB) reactor is shown in Figure 3-5 (the reactor to the left). Char exits the CFB through the cyclone separator and then goes into the char combustor that then heats a mixture of char and sand. Heat is transferred to the CFB utilizing a sand medium that is circulated between the char combustor and the CFB. Steam is provided at the bottom of the CFB reactor in order to allow the necessary chemical reactions to take place with the biomass.

The circulating & bubbling fluidized bed gasifier reactors

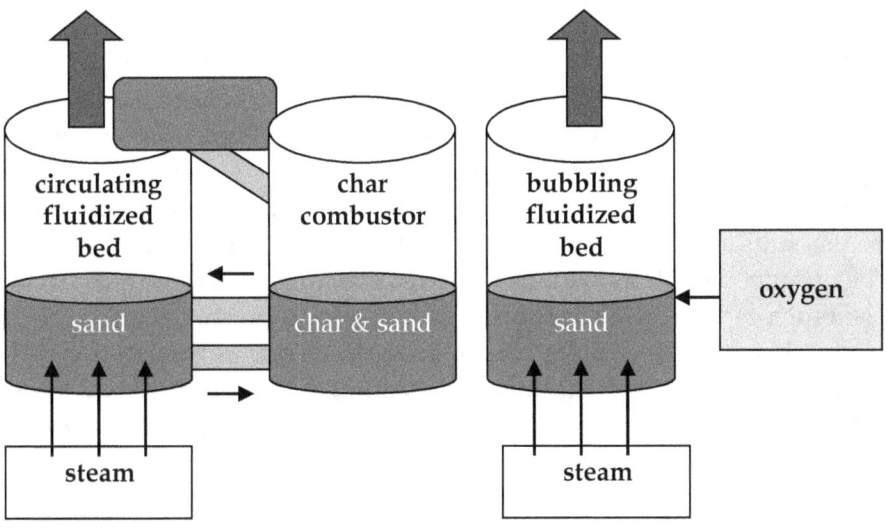

Figure 3-5 : Two types of fluidized bed gasifier reactors utilize sand as an effective heat exchange for gasification reactions. The sand is circulated between a char combustor and the reactor. The bubbling fluidized bed keeps the sand and fluidizing agents such as steam and oxygen within the reactor.

The bubbling fluidized bed reactor (to the right in Figure 3-5) operates differently in that it does not contain a char combustor next to it. This reactor employs a sand bed mixed with fluidizing agent(s) such as a steam-oxygen mixture that becomes inter-dispersed throughout the bed. Fluidized beds can also operate with the input of high pressure oxygen, sometimes mixed with steam by implementing an Air Separation Unit (ASU) where oxygen is extracted from the air via a cryogenic mechanism.[114]

When thermochemical gasification employs a mixture of biomass and coal, a reactor called an **entrained flow gasifier** is utilized. This gasifier works by inputting the coal/biomass as well as steam, air, etc at the top of the reactor. The leftover waste called slag exits from the bottom of the reactor. Other companies are also developing thermal gasifiers that vary from the basic fluidized bed design. One example is the **pulse combustion gasifier** built by Thermochem Recovery International.[115] This gasifier model is utilized by several companies involved with the integrated biorefinery demonstration program that manufacture diesel fuel. The pulse combustion gasifier operates on the principles of both pyrolysis and thermochemical gasification.

3.5 The essential process of cleaning synthesis gas requires several pieces of equipment

An essential process required to produce renewable diesel from thermal gasification is the removal of impurities from the synthesis gas stream. The gas cleaning process requires specialized equipment that makes up a sizeable portion of the overall refinery capital costs. It is a technically challenging process that removes impurities such as tars, char, soot, alkali minerals, hydrogen sulfide and ammonia. In addition, coal sources containing sulfur and heavy metal content such as mercury must also be removed. Therefore, the cleaning of synthesis gas requires several pieces of equipment as shown in Figure 3-6 below. The cleaning equipment described in this section is based upon the NREL indirect thermal gasification model that produces mixed alcohols & ethanol and the production of synthesis gas made from coal sources.[116,117]

Types of equipment for cleaning synthesis gas

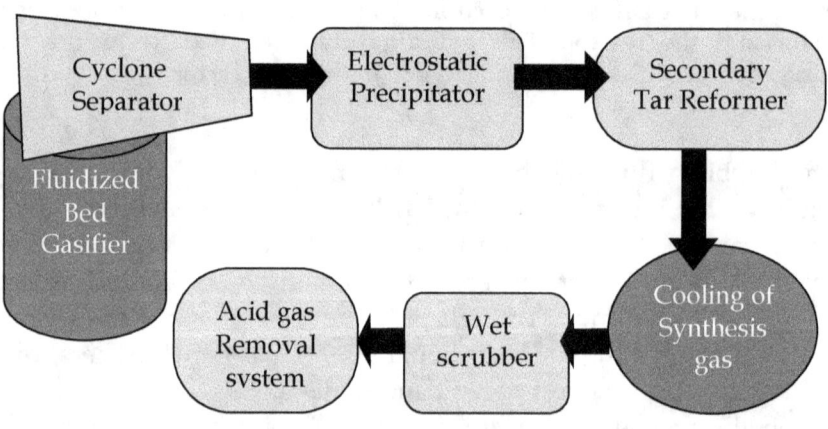

Figure 3-6 : Several types of equipment are utilized for the removal of particulates, tars, other contaminant and general cleaning of synthesis gas gases generated from biomass. The set of equipment shown here is based upon the NREL indirect gasification of mixed alcohols and ethanol along with synthesis gas made from coal.[116,117]

Figure 3-6 above outlines the type of cleaning equipment utilized in a thermochemical gasification refinery that generates synthesis gas from biomass sources. Basic cleaning equipment includes a **cyclone separator,** electrostatic precipitator, tar reformer, wet scrubber and an acid gas removal (AGR) system. The initial cleaning equipment required to remove particulates from the synthesis gas stream are the cyclone separator and an electrostatic precipitator located next to or near the gasifier. These pieces of equipment are effective at removing most types of particles from the gas stream leaving the gasifier. The rule of thumb is that the cyclone separator takes away particles over 5 microns while the electrostatic precipitator removes those that are smaller in size.

After this initial cleaning step, synthesis gas gets sent through a secondary tar reformer that thermally breaks down the tar compounds into further production gases. The gas then becomes cooled down and goes through further cleaning where a wet scrubber removes the remaining particles, as well as clears away the condensed ammonia and tars. In addition, another downstream process that consists of an acid gas removal (AGR) system has the function of removing hydrogen sulfide (H_2S) and carbonyl sulfide (COS) from the synthesis gas so it will not adversely affect the operation of the metal catalysts in the synthesis stage.

3.6 The thermochemical processing method of pyrolysis produces a bio-oil that must be upgraded in order to be utilized as a vehicle fuel

Pyrolysis is another thermochemical process where heated biomass forms organic vapor compounds that undergo a rapid exposure time and then become rapidly cooled in order to form a condensed bio-oil. The combined thermal processes forming the bio-oil originating from the pyrolysis reactor are shown in Figure 3-7. The heating of the biomass takes place within a reactor in the absence of air or oxygen. **Volatile organics** are the immediate compounds formed in the pyrolysis reactor during the heating process. The volatiles then go through a short holding period called **residence time**. After the residence time period the organic end product compounds get rapidly quenched into bio-oil. The allotted residence time determines the type of pyrolysis method employed. Pyrolysis can be classified as slow, fast or flash. The specific ranges of residence time, temperature range and heating value of the types of pyrolysis are outlined in Table 3-2.

The reaction conditions for pyrolysis are conducted at temperatures between 300 to 950 ° C while the volatiles residence times range from 0.5-2 seconds up to 30 minutes.[118,119] **Flash pyrolysis** has residence times that take place under 1 second while **fast pyrolysis** times take place up to several seconds and **slow pyrolysis** has residence times that range from 5 to 30 minutes.[119] Also, pyrolysis varies according to the heat ramping rates that take place within the reactor. Heat ramping rates vary from several degrees per minute up to greater than one thousand degrees per minute.[120]

Fast and slow pyrolysis differ in the relative amounts of bio-oil and char they produce. Fast pyrolysis tends to make more amounts of the bio-oil and less char when compared to slow pyrolysis. In fact, slow pyrolysis at times tends to combine the char and bio-oil together in what is known as **bioslurry**. Bioslurry must be handled differently when compared to straight bio-oil.

Pyrolysis process diagram outlining the steps that create bio-oil

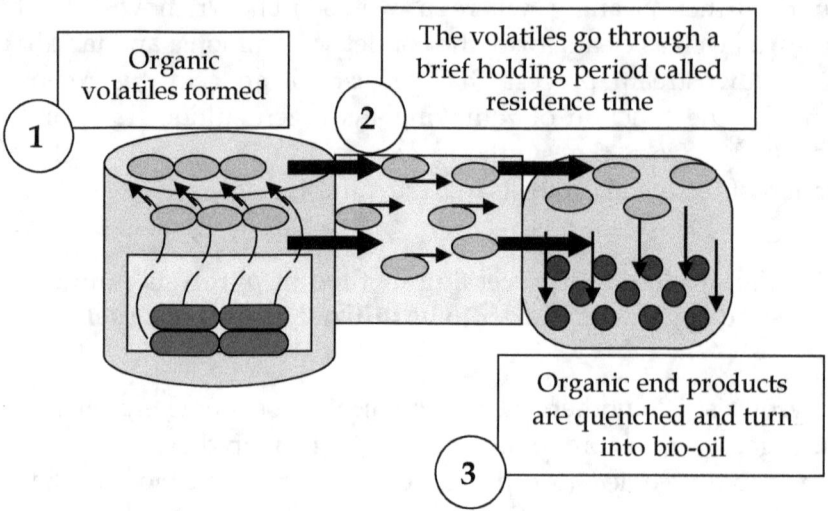

Figure 3-7 : Pyrolysis involves several reaction phases. Initial organic volatiles are formed in the reactor during the heating ramp up period. Afterwards, the organic volatiles go through an intermediate holding time period called residence time. Subsequently the organic end products are rapidly quenched to form bio-oil.

The three main products made from the pyrolysis of biomass include **bio-oil**, **char** and **producer gas** as shown in Figure 3-8. During the production process the gases, organic volatiles and char are emitted from the reactor after initial heating and subsequent release. A cyclone unit separates the char from the other components. The remaining gases and organic volatiles travel to a quenching unit that condenses the organics into a bio-oil. The producer gas is directed to other parts of the refinery for further energy production or alternative applications. Producer gas consists of synthesis gas components as well as other light weight hydrocarbons.

Classification of pyrolysis types based upon residence times, Temperature range and heating rates

Pyrolysis type	Heating rates	Temperature range	Gas residence times	Determination of products
Flash, Fast or Slow	2 - 2000 °C per minute	300 - 950 °C	<1 or 2 seconds 5 - 30 minutes	Fast pyrolysis produces more bio-oil than char

Table 3-2 : Flash, fast or slow pyrolysis can be classified according to heating rates, temperature range, gas residence time and types of bio-oil & char products.[118,119,120]

Schematic of reaction products and processes associated with pyrolysis

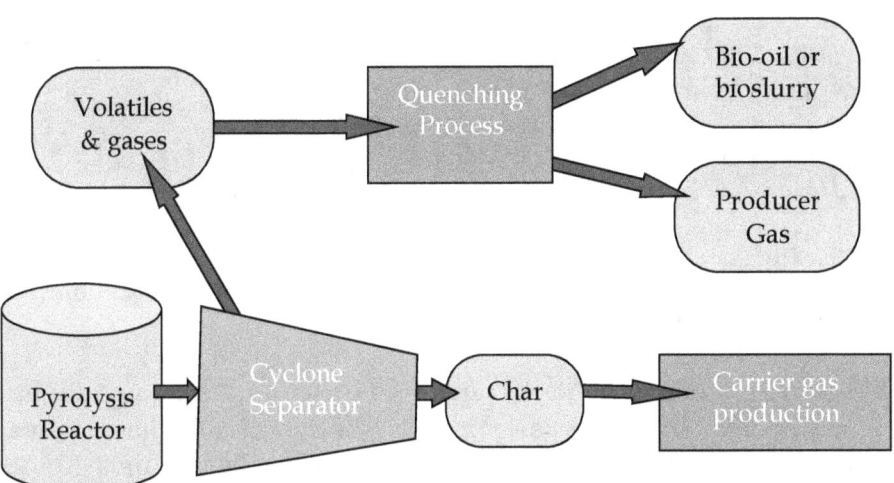

Figure 3-8 : The pyrolysis process involves the production of volatile organics, char or gases where the volatiles are condensed into bio-oil and the gases get utilized as producer gas. The char is initially separated from the organic volatiles and gases by a cyclone separation unit while a quenching unit helps form the bio-oil from the volatile organics. The char itself can be implemented towards carrier gas production.

Bio-oil is the main product of interest obtained from pyrolysis and requires an upgrading process in order for it to be utilized as vehicle fuel. Bio-oil consists of a complex mixture of 300 or more individual types of organic compounds that can be ketones, aldehydes, carboxylic acids, sugars, phenolics or other types.[121] This bio-oil contains both water and hydrocarbon solvent soluble portions. The hydrocarbon solvent, utilized for the extraction or separation of certain organics in the bio-oil, may normally be a non-polar type of solvent. However, these different bio-oil portions can also be separated from one another by simply adding water and then decanting the mixture.

The water solubility of bio-oil is correlated to its oxygen content. If the oil has higher oxygen content, more types of compounds are usually soluble in the water (aqueous) portion. Figure 3-9 shows the types of compounds contained in bio-oil and their relative solubility (or insolubility) in water. Polysaccharides, alcohols and organic acids are some of the water soluble components of bio-oil. The hydrocarbon solvent soluble portion contains compounds like lignins and aromatics. Compounds such as esters, ethers, aldehydes and ketones can be soluble in either the water or hydrocarbon solvent soluble portions depending upon their size and structure.

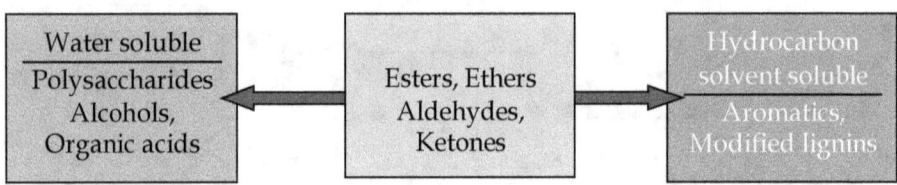

Figure 3-9 : The water soluble or hydrocarbon solvent soluble fractions of bio-oil. Certain compounds can be classified as either water soluble or insoluble while other organics could be separated into either fraction.

Bio-oil as a product, by itself, cannot be directly used as a ready made vehicle fuel source. A high oxygen content in the bio-oil compounds prevents its application as vehicle fuel. These compounds normally contain a 30 - 40 % oxygen content.[122] The high oxygen content bio-oil has adverse properties such as high viscosity, corrosivity and poor thermal stability associated with it. This becomes evident upon examination of the bio-oil values of pH, heating value and viscosity.[123] The bio-oil normally has pH values from 2-3 which makes it too corrosive. It contains a heating value of around 17 MJ/kg which denotes it has too low of an energy content. It also has a viscosity value that often ranges from 40 – 100 mm/s making it too high for good flow characteristics.

Due to these adverse characteristics, the raw bio-oil must be post processed into the refined organic compounds that do not have poor chemical and physical properties. Post processing methods involving catalytically based upgrading are effective in converting the bio-oil into viable biofuel. The catalytic upgrading methods of hydrocracking and hydrotreating allow fuels such as renewable gasoline and diesel to be produced.

3.7 Rapid thermal processing (RTP) is a type of pyrolysis that produces a variety of vehicle fuels

Figure 3-10 below demonstrates a method that makes a combination of fuels such as gasoline, kerosene and diesel from a flash pyrolysis technology known as **rapid thermal processing (RTP)**. The pyrolysis reactor mainly produces a combination of producer gas and bio-oil. After it has undergone filtration and separation, as shown in the figure, the bio-oil then becomes post processed through a hydroprocessing unit in order to form usable vehicle fuel. Keep in mind that the hydroprocessing unit requires an adequate supply of hydrogen in order to carry out the hydrocracking and hydrotreatment processes.

The UOP/Ensyn Rapid Thermal Processing (RTP) pyrolysis manufacturing technology that produces a variety of fuels

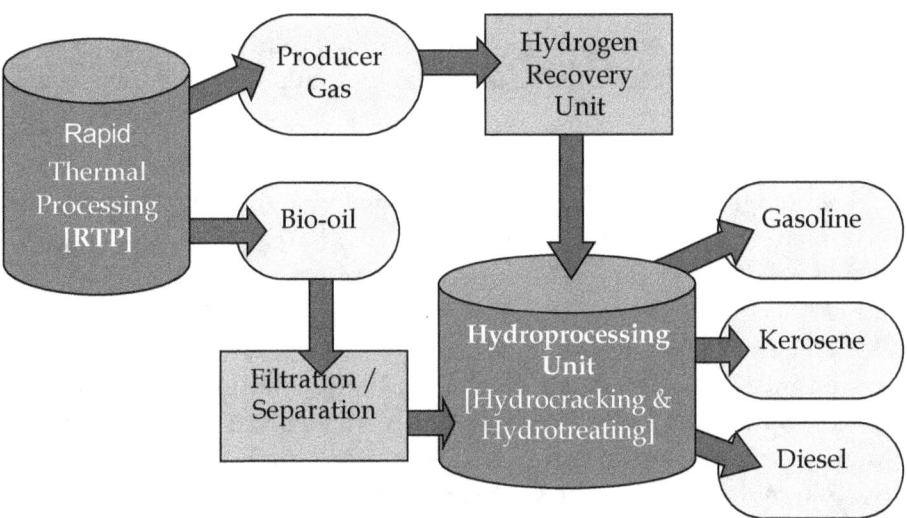

Figure 3-10 : The Rapid Thermal Processing method processes a number of different biomass feedstocks that are eventually converted into a number of potential fuels such as gasoline, kerosene and diesel accomplished through hydroprocessing technology combining hydrocracking and hydrotreating.[124]

They obtain this hydrogen gas supply by processing producer gas through a hydrogen recovery unit. The hydrocracking and hydrotreatment processes tied together modify the many organics in the bio-oil into a variety of usable fuel compounds. They do this by recombining organic compound groups and reducing the amount of oxygen contained in the bio-oil compounds.

Rapid thermal processing has already proven to be successful on a large production scale. The fact that this method produces a higher amount of bio-oil and lower amounts of char and biogases makes it an attractive biofuel refinery model to implement. The amount of char and gases produced from RTP are low, ranging from 10 – 15 % of the total products, each respectively.[125] Another advantage of RTP is that it that can incorporate a wide variety of feedstock sources such as forestry wastes, agricultural residues, paper, wood, municipal wastes and grasses in order to make alternate fuels. The company UOP utilizes this technology for the purpose of producing green gasoline and other fuels [ie jet fuel].[124] They operate a pilot demonstration plant in Hawaii that has qualified as an award recipient in the Integrated Biorefinery demonstration program. The refinery claims to be able to produce 4 barrels of fuel (i.e., > 120 gallons) for every 1 ton of feedstock input into it. They also have made a number of cooperative purchasing agreements in order to obtain a variety of feedstocks that include corn stover, cane bagasse, switchgrass, gunieau grass, algae biomass and paper waste.[124]

3.8 Municipal waste, energy crops and residues require milling in order to manufacture biofuel from thermochemical methods

Compared to other types of biofuel manufacture, all the thermochemical based technologies have the advantage of being able to utilize a wide range of biomass sources for conversion into biofuel. Therefore, a large number of renewable feedstocks can be recovered from various industries in large quantities. According to the USDA & DOE billion ton study, over 1 billion tons of renewable crops and/or waste sources are available towards conversion into biofuel or other products as stipulated by certain cultivation or recovery options.[126] The majority of this biomass consists of crop or forestry residues as well as energy crops. In addition, municipal waste accounts for another 200 million tons or more of waste available in the United States each year.

Therefore, the allocation of renewable resources eligible for biofuel manufacture can be classified into three categories as shown below in Table 3-3. These categories include energy crops, municipal waste and renewable residues. Popular types of energy crops include switchgrass, miscanthus, hybrid poplar, aspen-birch and also algae & peat moss. Renewable residues include agricultural crop and forestry related wastes. Some of the agricultural residues available consist of corn stover, wheat-barley-rice straw & sugarcane/sorghum bagasse. Forestry residues can be obtained from mills, logging wastes, electrical energy generation and forest fire timber to name a few. Municipal wastes include food scraps, paper & cardboard, wood wastes, plastics, demolition & construction wastes and also wastewater sludge. Cellulose or wood derived renewable wastes or materials are a part of all three classified categories.

The categories of renewables that can be utilized for biofuels production

Energy crops	Municipal waste	Renewable residues
Switchgrass, algae, miscanthus, peat moss, hybrid poplar, aspen-birch	Paper & cardboard, plastics, food/yard scraps, demolition & construction wastes, wastewater sludge	Lumber mills, forest fire timber, logging residues, Corn stover, wheat-barley-rice straw, sugarcane & sorghum bagasse

Table 3-3 : A large variety of renewable sources or wastes are available for manufacture into biofuels. They can be classified into one of three categories which are energy crops, municipal waste and renewable residues.

Biofuel manufacture tends to take place utilizing a single feedstock source or a combination of them. For example, multiple feedstock sources input into a biorefinery can include coal and switchgrass for thermochemical gasification, and crop residues, grass and paper waste for pyrolysis. Examples of single feedstock sources utilized in thermochemical refineries include wood chips input into a pyrolysis unit; algae utilized in a hydrotreatment facility; and municipal waste incorporated into a thermochemical gasification refinery. Municipal waste is ideal for thermochemical gasification due to the newer plasma melting technology, covered later in chapter 5.

A primary factor that affects the availability of the renewable feedstocks chosen for a refinery is the relative location of these resources in relation to the refinery itself. These resources should be preferably located near the refinery. Another primary factor that affects the choice of renewable feedstocks utilized in a refinery is the pre-processing methods necessary to size down the material appropriate for use in the thermochemical reactor. To successfully synthesize fuel compounds, various pre-processing methods are essential to breakdown the biomass to an acceptable form. The pre-processing steps implemented for feedstock conversion are shown in Figure 3-11.

In the majority of cases, feedstock sources require primary and secondary pre-processing steps which are executed in biomass conversion facilities or mills. Primary and secondary milling can also be done within the same facility or performed in separate ones where the material is transferred from one facility to another. For example, perennial woods are initially cut into wood chips before they are transferred to another milling facility that further shreds and reduces them in size. Switchgrass goes through several consecutive milling steps after it is delivered as bales to the appropriate conversion facility. In other cases, forestry type wastes like sawdust and wood shavings recovered at sawmills would not require as much processing and could be transported directly to the refinery.

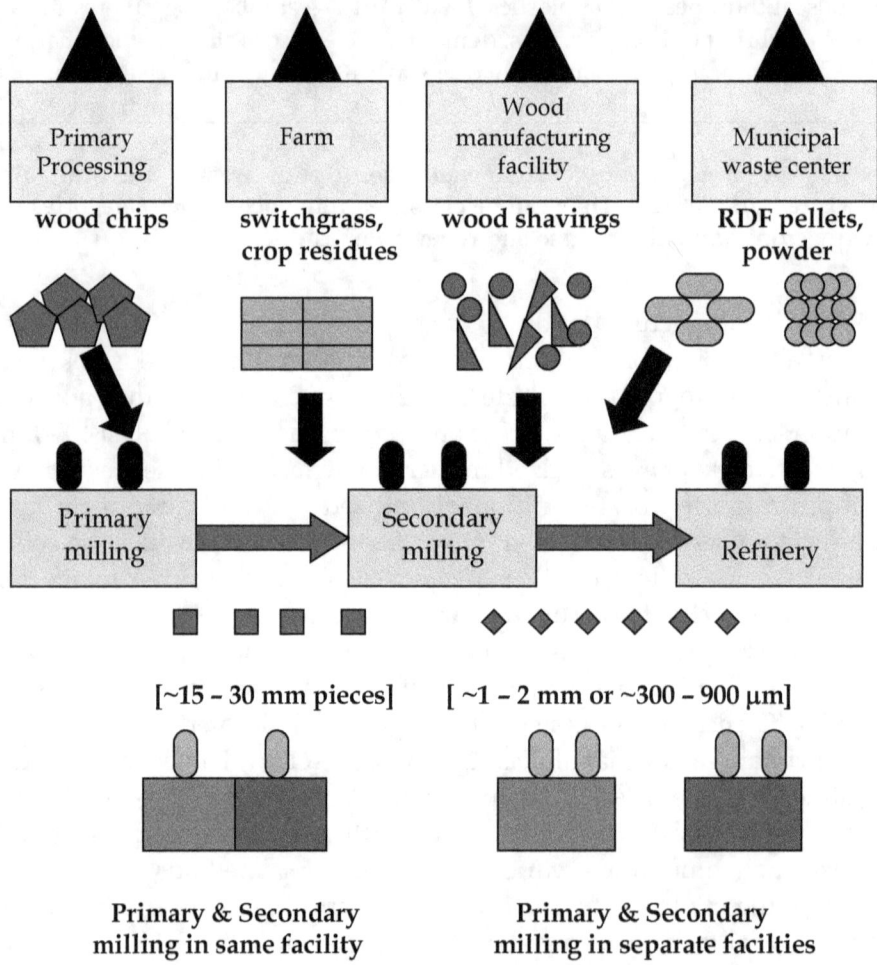

Figure 3-11 : Renewable feedstocks such as wood chips, agricultural residues & switchgrass can require both primary & secondary milling before being delivered to a refinery, whereas feedstocks such as wood shavings or municipal waste may only require one processing step before utilized in refineries.

Municipal waste goes through several processing stages before it is converted into powder or pellets. This material can then be transferred to the appropriate refinery or waste conversion facility. Oftentimes this municipal waste gets delivered to Waste to Energy (WTE) facilities but could just as easily be transported to vehicle fuel production refineries if they were already set up.

Feedstock preparation at biomass conversion facilities involves cutting and/or grinding the material with a hammer, ball or knife mill. Thereafter, sieving and screening processes are performed until the material gets sorted into the proper size range. The milling and sieving processes can be repeated several times within the mill until the particle size is small enough for utilization. Lower particle sized material correlates to a larger amount of biofuel made and less amounts of char and gases produced.[128] Particle sizes optimal for biorefinery application range from ten millimeters down to hundreds of microns.[127] In most cases a particle size of several millimeters or less is sufficient for subsequent biofuel production. After milling is completed the biomass oftentimes requires drying in order to reduce its moisture content. Before being input into the thermal reactor the biomass must reach a final moisture content of 5 - 10 %. In summary, various pre-processing methods are essential towards providing the biomass for a refinery in the acceptable form and consistency in order to successfully synthesize fuel compounds using thermochemical refinery technologies.

3.9 Thermochemical liquefaction has some basic similarities and differences compared to pyrolysis and can be performed on an industrial scale

Another thermochemical processing method known as **liquefaction** produces a sizeable amount of bio-oil that can be later converted into vehicle fuel similar to what is done with pyrolysis. Liquefaction takes place under conditions of moderately high temperature and pressure in a liquid medium. Liquefaction is executed at temperatures usually ranging from 100 to 500 ° C, with pressures from 5 to 30 MPa [megapascals] and reaction periods from 15 to 60 minutes in duration as is shown in Figure 3-12.[129] The initial biomass concentration contained in the liquid medium usually ranges from 10 to 40 %.[130] When the medium happens to be water the process is called hydrothermal liquefaction. Similar to pyrolysis, the bio-oil contains both water soluble and hydrocarbon solvent soluble fractions.

A quality bio-oil is made during liquefaction due to the addition of key components that get introduced into the production system. These additional materials can include carrier gases and mineral catalysts. Oftentimes the carrier gas consists of synthesis gas components that are fed into the reactor. The delivery and placement of the hydrogen and carbon monoxide gases into the reactor greatly assists towards producing a high quality bio-oil. In addition, other beneficial reactions can take place with liquefaction due to the addition of a mineral based catalyst like sodium carbonate that gets placed into the bioslurry mixture at around a 1 - 5 % concentration.[129]

The process conditions carried out with liquefaction

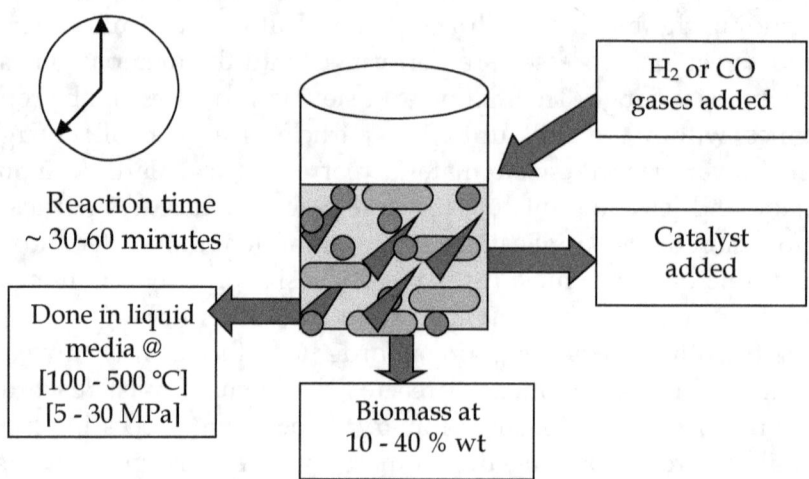

Figure 3-12 : Liquefaction is performed with a liquid medium containg biomass added at 10 - 40 %, a reducing gas such as hydrogen or carbon monoxide and oftentimes uses a catalyst. It is carried out at temperatures from 100 - 500 °, pressures of 5 - 30 MPa and reaction times usually ranging from 30 – 60 minutes.[129,130]

An industrial based liquefaction production scheme is shown in Figure 3-13.[131] The type of production system is mainly a design idea and therefore may not be practiced by companies in general as of yet. The liquefaction process takes place in a reactor containing water as the medium. The initial bio-oil product made is called the post reacted slurry. Further treatment of the post-reacted slurry takes place by executing certain solid/liquid separation methods like decantation, filtering or hydrocarbon based extraction. The separation process tends to make three portions; a hydrocarbon based bio-oil, an aqueous based organic fraction and a solids/char portion. The industrial liquefaction method also yields gaseous products and vapor condensate fractions. The gaseous products can be further separated and gathered as carbon dioxide & small hydrocarbons or as a synthesis gas stream that gets recycled back into the reactor. The condensate portion consists of volatile organic carbons collected through condensation much like what takes place with bio-oil formed from pyrolysis.

There are also several advantages of performing liquefaction over pyrolysis. These advantages are outlined in Table 3-4. They include reduced oxygen content in the bio-oil and a higher inherent moisture content contained in the feedstock that is utilized. Of primary importance is the reduced amount of oxygen contained in the bio-oil. Table 3-4 shows the relative amounts of carbon and oxygen found in the bio-oil from both pyrolysis and liquefaction. Pyrolysis bio-oil usually consists of organic compounds having a similar percentage distribution of carbon and oxygen ranging from 40 to 55 percent.[132] Conversely, liquefaction based bio-oil has a much higher carbon content that is above 70 percent while the oxygen content usually ranges between 10 – 20 percent.[132] Higher carbon content in the bio-oil denotes that the bio-oil is more easily fashionable into vehicle fuel since it would require a minimal amount of additional post processing.

Outline of unit processes and end products resulting from industrial based liquefaction

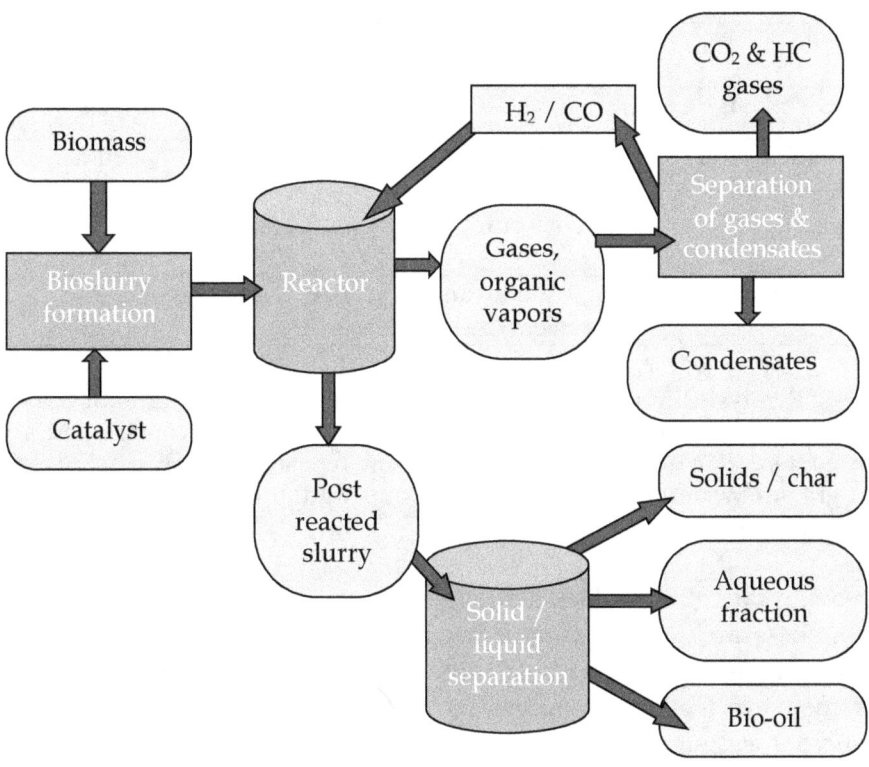

Figure 3-13 : Liquefaction produces a post-reacted bioslurry made from the biomass and catalyst reacting with synthesis gas (H_2/CO). Solid/liquid extraction then takes place where the bioslurry is separated into the three components of solids/char, an aqueous fraction and bio-oil. The gases and vapors are split up into condensates, carbon dioxide & hydrocarbon gases as well as the generation of synthesis gas.[131]

Performing liquefaction with a high moisture content biomass source is another advantage related to this particular thermochemical method. Biomass feedstocks utilized for liquefaction are allowed to contain higher moisture content materials as is demonstrated in Table 3-4. On the other hand pyrolysis requires feedstocks that have a much lower moisture content that usually ranges between 5 - 10 % while liquefaction implements biomass that normally contains at least a 50 – 60 % moisture content.[133] Therefore, liquefaction is better suited to handle most types of biomass since the method can theoretically utilize both low and high moisture content based feedstocks. This also denotes that an energy intensive process such as drying is not required to reduce the moisture content of the biomass sample in order to conduct liquefaction.

Carbon & oxygen percentage of bio-oil along with the original moisture content of the feedstock for liquefaction and pyrolysis

	Carbon %	Oxygen %	Moisture Content
Pyrolysis	Around 40 % +	> 50 %	10 – 25 %
Liquefaction	70 - 75	10 – 20	> 50 %
Types of high moisture feedstocks			
Sludge, black liquor, water hyancith, grasses, microalgae, manure, brewery/refinery waste, kelp			

Table 3-4 : Pyrolysis bio-oil has an even percentage distribution of both carbon and oxygen while liquefaction bio-oil has a much higher carbon content and a lower oxygen content.[132] Also, high moisture content feedstocks such as sludge, black liquor, grasses, water hyancith, manure, kelp and brewery refinery waste are suitable for liquefaction.

A wide range of feedstock sources are available for liquefaction. While some of these are sources typical for pyrolysis, others have naturally high moisture content that make them more suited to liquefaction. Some of these higher moisture feedstock sources include sewage sludge, black liquor/tall oil, micro or macroalgae, peat, water hyancith, kelp, manure and grasses.[134,135,136,137,138,139,140] Liquefaction has also been executed with lignocellulosic sources such as ground up wood chips or waste newsprint where the process has been performed in a liquid oil media instead of water. In the past, small scale liquefaction based experiments have also been executed by U.S. government energy laboratories utilizing a variety of feedstock sources.[139] As part of the established experiments, high moisture content feedstocks such as macrocystis kelp, napier grass, water hyancith and spent grain from brewery waste were used.

However, a disadvantage with liquefaction, as well as with pyrolysis, is that the bio-oil produced contains nitrogen based hydrocarbon and heterocyclic compounds such as pyridines, pyrolles, pyrazines, quinolines, indoles, amides, amines, etc.[141] This is a major drawback since an environmentally friendly biofuel should contain little to no nitrogen or sulfur based compounds. However, a subsequent hydrotreatment process can remove the nitrogen in these organics by emitting it as ammonia.

In addition, there is the potential of utilizing another thermochemically based production method that would altogether prevent the formation of nitrogenous compounds caused by the breakdown of proteins. This method relates to the formation of a biocrude from algae or microbes, which is discussed in the next section.

3.10 A two step liquefaction process creates a biocrude from algae

Certain liquefaction techniques from sources like algae can produce either a biocrude or a bio-oil. The use of both terms is oftentimes synonymous, but the differences mainly apply to which end products are made from them. Bio-oil consists of hundred(s) of different compounds that are eventually converted into biofuel as previously discussed earlier in the chapter. On the other hand, a useful biocrude can consist of a set of individual compound units that have the potential to be separable from one another. These units are composed of distinct sets of just amino acids or saccharides or fatty acids. Algae itself contains the basic biomolecular materials of proteins, polysaccharides and lipids which can be broken down into these individual units.

Biocrude made from algae has the potential to be converted into biofuel and component chemical ingredients while bio-oil from liquefaction is normally slated just for biofuel production. This is due to the fact that the individualized sets of compounds contained in biocrude are more fashionable towards making ingredients for certain products. Such examples include certain fatty acids that are input into consumer health and hygiene products. Amino acid components make for a healthier form of human or animal feed supplement. Saccharides can be further fashioned into chemicals or converted into biofuel compounds.

Another basic difference between bio-oil and biocrude production has to do with how nitrogen based compounds are handled. For bio-oil, these compounds can be a hindrance towards producing a quality biofuel. On the other hand, the peptide or amino acid content contained in biocrude can be conceivably separated from the overall mixture before biofuel production takes place. A method for creating a separable amino acid hydrolysate from a biocrude formation process is described below. Such a process consists of a two step hot water extraction and enzyme based hydrolysis method as shown in Figure 3-14.

A two step hot water and enzymatic hydrolysis process

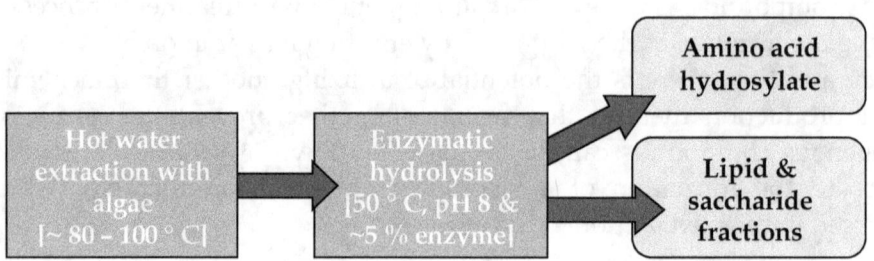

Figure 3-14 : Algae can effectively be separated into an amino acid hydrosylate and a lipid/saccharide fraction through a two step extraction process. The two step process consists of a hot water extraction done at temperatures around 80 - 100 ° C and then a resulting enzymatic hydrolysis method at conditions of 50 ° C, pH 8 and a 5 % enzyme concentration.[142]

The initial hot water extraction process includes heating the algae sample at around 80 ° C while the subsequent enzymatic hydrolysis part involves the addition of a 5 % concentration of enzyme at pH 8 which is heated at around 50 ° C.[142] The separated amino acid hydrosylate can serve as a rich, valuable source for bio-based fertilizer, animal feed or a healthy human nutritional consumptive product. The remaining lipids and saccharides portion are then ready to be converted into further biofuel or become separated into their respective components utilized as ingredients in products.

REFERENCES:

110. **Introduction to Fischer Tropsch technology** – Studies in Surface Science and Catalysts Vol 152 pgs 1 – 63 [2004] by AP Steynberg
111. **Fischer Tropsch fuels refinery design** – Energy and Environmental Science Vol 4 pgs 1177 – 1205 [2011] by A. de Klerk
112. **The Fischer-Tropsch Process [1950 – 2000]** – Catalysis Today Vol 71 pgs [2002] by M. Dry
113. **Thermochemical biomass gasification : A review of the current status of the technology** – Energies Vol 2 No 3 pgs 556 – 581 [2009] by A. Kumar, DD Jones, MA Hanna

114. **Gasifipedia supporting technologies – air separation –** National Energy Technology Laboratory (NETL) – http://www.netl.doe.gov/technologies/...
115. **TRI Technology – Steam reforming gasification process –** http://www.tri-inc.net/TRI-inc/Technology.html -- accessed May 2014
116. **Thermochemical ethanol via indirect gasification and mixed alcohol synthesis of lignocellulosic biomass –** NREL (National Renewable Energy Laboratory) – [Apr 2007] by S. Phillips, A. Aden, J. Jechura, D. Dayton
117. **Affordable, low carbon diesel fuel from domestic coal and biomass –** DOE/NETL 2009/1349 [Jan 2009] by TJ Tarka
118. **Application of a particle model to pyrolysis. Comparison of different feedstock : Plastic, tyre, coal and biomass –** Fuel Processing Technology Vol 103 pgs 1 – 8 [2012] by MV Navarro, JD Martinez, R Murillo, T. Garcia et al
119. **Bench scale fluidized bed pyrolysis of switchgrass for bio-oil production –** Industrial and Engineering Chemistry Research Vol 46 pgs 1891 – 1897 [2007] by AA Boateng, DD Daugaard, NM Goldberg, KB Hicks
120. **Thermal decomposition of bio-oil : Focus on the products yields under different pyrolysis conditions –** Fuel Vol 102 pgs 274 – 281 [2012] by Y. Chhiti, S. Salvador, JM Commandre, F. Broust
121. **Pyrolysis of wood/biomass for bio-oil : A critical review –** Energy and Fuels Vol 20 pgs 848 – 889 [2006] by D. Mohan, CU Pittman, PH Steele
122. **A review of catalytic upgrading of bio-oil to engine fuels –** Applied Catalysis A : General – Vol 407 pgs 1 – 19 [2011] by PM Mortensen, JD Grunwaldt, PA Jensen et al
123. **Review of fast pyrolysis of biomass and product upgrading –** Biomass and Bioenergy Vol 38 SI pgs 68 – 94 [2012] by AV Bridgwater
124. **Pilot scale biorefinery for sustainable fuels from biomass via integrated pyrolysis and catalytic conversion – Integrated biorefinery (IBR) project update –** EU Biomass Conference – IEA Bioenergy Task 42, Biorefining – 8th task meeting – Chicago, Oct 2010 by S. Lupton of UOP LLC
125. **Commercial scale rapid thermal processing of biomass –** Biomass and Bioenergy Vol 7 pgs 251 – 258 [1994] by RG Graham, B Freel, D Huffman et al
126. **Biomass as a feedstock for a bioenergy and bioproducts industry : The technical feasibility of a billion ton annual supply – Factors increasing biomass resources from agriculture –** by US Department of Energy & US Department of Agriculture – R. Perlack, L. Wright, A. Turhollow, R.Graham, B.Stokes et al [Apr 2005]
127. **Oxidative lime pretreatment of Alamo switchgrass –** Applied Biochemistry and Biotechnology Vol 165 No 2 pgs 506 – 522 [2011] by M. Falls, MT Holtzapple
128. **Effect of pretreatment on the physical properties of biomass and its relation to fluidized bed gasification –** Environmental Progress and Sustainable Energy Vol 31 No 3 pgs 335 – 339 [2012] by B. Bronson, F. Preto,

129. **Non catalytic and catalytic hydrothermal liquefaction of biomass** – Research on Chemical Intermediates Vol 39 No 2 pgs 485 – 498 [2013] by K. Tekin, S. Karagoz

130. **Direct liquefaction of lignocellulosic residues for liquid fuel production** – Fuel Vol 94 pgs 324 – 332 [2012] by S. Bensaid, R. Conti, D. Fino

131. **Biomass liquefaction : An overview** in **Fundamentals of Thermochemical Biomass Conversion** pgs 967 – 1002 [1985] (Springer Netherlands) by E. Chornet, RP Overend

132. **Renewable fuels via catalytic hydrodeoxygenation** – Applied Catalysis A : General Vol 397 pgs 1 – 12 [2011] by TV Choudhary, CB Phillips

133. **A review on operating parameters for optimum liquid oil yield in biomass pyrolysis** – Applied Catalysis A : General Vol 397 pgs 1 – 12 [2011] by TV Choudhary, CB Phillips

134. **Catalytic hydrotreating of black liquor oils** – Energy and Fuels Vol 5 pgs 102 – 109 [1991] by DC Elliot, A Oasmaa

135. **Characterization of peat and biomass liquids** in **Fundamentals of Thermochemical Biomass Conversion** (Elsevier Applied Science Publishers) – [1985] pgs 1019 – 1026 by D. Karlsson, D. Bjornbom

136. **Production of heavy oil from sewage sludge by direct thermochemical liquefaction** – Desalination Vol 98 pgs 127 – 133 [1994] by S. Itoh, A. Suzuki, T. Nakamura, SY Yokoyama

137. **Hydrothermal liquefaction of Nannochloropsis sp. Systematic study of process variables and analysis of the product fractions** – Biomass and Bioenergy Vol 46 SI pgs 317 – 331 [2012] by PJ Valdez, MC Nelson et al

138. **Hydrothermal liquefaction of separated dairy manure for production of bio-oils with simultaneous waste treatment** – Bioresource Technology Vol 107 pgs 456 – 463 [2012] by CS Theegala, JS Midgett

139. **Product analysis from direct liquefaction of several high moisture biomass feedstocks** in **Pyrolysis Oils from Biomass** (ACS Symposium Series Vol 376) - chapter 17 pgs 179 – 188 [1988] by DC Elliot, LJ Sealock, RS Butner

140. **Hydrothermal processing of macroalgae feedstocks in continuous flow reactors** – ACS Sustainable Chemistry and Engineering Vol 2 No 2 pgs 207 – 215 [2014] by DC Elliot, TR Hart, G Neuenschwander et al

141. **Fractionation and identification of organic nitrogen species from bio-oil produced by fast pyrolysis of sewage sludge** – Bioresource Technology Vol 101 pgs 7648 – 7652 [2010] by JP Cao, XY Zhao, K Morishita, XY Wei, T. Takarada

142. **Development of a process for the production of L-amino acids concentrates from microalgae by enzymatic hydrolysis** – Bioresource Technology Vol 112 pgs 164 – 170 [2012] by JMR Garcia, FGA Fernandez, JMF Sculla

Chapter 4

Biodiesel Manufacture

4.1 Biodiesel production still has a potential place in our biofuel infrastructure

Biodiesel mainly has a place in our current fuel infrastructure as an additive to petrodiesel for the purpose of improving certain properties in it. In addition, biodiesel tends to be cleaner and more environmentally friendlier than petrodiesel. However, some disadvantages make pure biodiesel (B100) difficult to implement as a stand-alone fuel in modern diesel engines. B100 cannot be utilized directly in diesel engines manufactured later than 2007 due to changes in engine design that affect the exhaust and lubrication systems. However, biodiesel still has the potential to be utilized in modern vehicle engines if certain changes were made to it. For example, the placement of certain fuel additives in it could correct the adverse properties associated with its use in diesel engines.

The manufacture of biodiesel also offers other advantages related to industry. In general, increased biodiesel production would tend to create other manufacturing or agricultural based industries. Since biodiesel is made from renewable resources there would be a need for increased cultivation of oilseeds and algae. In addition, the generation of waste glycerine creates a chemical commodity useful as a potential ingredient in a plethora of products. Glycerol has a bright future in the chemical and other product based industries.

To fully grasp biodiesel manufacturing processes, it is helpful to understand the behavior and properties that the fuel imparts under certain circumstances. Many properties of biodiesel are derived from the specific oilseed sources from which it is made. This chapter discusses how these oilseed sources affect the specific properties of biodiesel. Certain oilseeds may also have advantages over other types. For example, desert oilseed sources offer more choices and capabilities towards producing biodiesel. Those interested in the industry should also comprehend material that involves improvements related to manufacturing done on a larger or smaller scale. Some of these processes include sonication or enzyme based production. Improvements such as these give biodiesel manufacture an important place in our biofuel infrastructure.

Learning the biodiesel production process requires knowledge of the chemical names & structures as well as the reactions that take place with the vegetable oil sources utilized to make it. Biodiesel consists of similar organic compounds made from the chemical modification of lipid precursor compounds known as triacylglycerides (TAGs). The TAGs get converted into organic compounds called alkyl esters which in essence make up biodiesel. The most common type of alkyl esters are methyl esters. The chemical structures of TAGs and methyl esters along with the required reactants necessary for biodiesel production are shown in Figure 4-1. Basically fatty acids from TAGs combine with a hydroxide catalyst and an alcohol (ethanol or methanol) in order to make an alkyl ester (a methyl ester shown in the figure).

Biodiesel alkyl esters are made from triacylglycerides (TAGs) combined with hydroxide and an alcohol

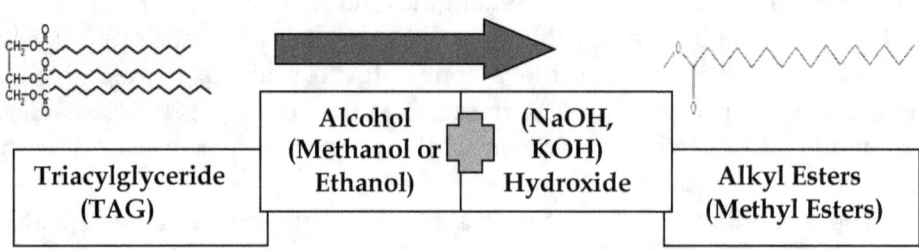

Figure 4-1 : Biodiesel consists of organic compounds called alkyl esters which are made by the chemical reaction that takes place with triacylglycerides combined with an alcohol and a hydroxide.

4.2 The main purpose for biodiesel production in the US is to mix it with petrodiesel in order to give it a higher lubricity

After ethanol, biodiesel is the second largest biofuel manufactured worldwide. It has been projected that around 37 billion gallons of biodiesel per year will be made across the world by 2016.[143] The US government has set the standard specifications for biodiesel blends in regular diesel fuel ranging from 5 % (B5) to 20 % (B20) as stipulated by the Energy Independence and Security Act of 2007. A 5 % blend of biodiesel in regular petroleum diesel would require around 3 billion gallons of biodiesel manufactured per year. The future stipulated production goal of 20 % biodiesel (B20) would need approximately 12 billion gallons of it produced per year.

Currently the US produces just over 1 billion gallons of biodiesel per year, or only about one-third of the amount required for the 5 % blend. However, since biodiesel is still manufactured at such a low quantity, the blend percentage of it mixed into petrodiesel varies across the country.

Therefore, it oftentimes differs from the advertised percentage shown at filling stations. Since three billion gallons of biodiesel is a large amount, our current supply of oil and fat feedstock sources may not be able to satisfy this need. Nevertheless, this amount could be produced by increasing the availability and supply of natural oil sources from certain oilseeds and algae strains. For example, certain algae can excrete single free fatty acids, which would be very ideal for biodiesel manufacture.

Biodiesel must also fulfill requirements from official standards before it can be mixed in with regular petrodiesel. Biodiesel mixed at 5 – 7 % must meet the ASTM D975 standard while blend percentages of 6 – 20 % go by the ASTM D7467 standard.[144] Requirements for these standards include minimum glycerol, sulfur content and acid value. Table 4-1 below shows a list of requirements for biodiesel meeting the European B100 EN 14214 standard, which oftentimes is also applied in the United States.[145]

Requirements of minimum contaminants and other properties for B100 meeting the EN 14214 standard

Cetane number	47 min.
Acid value	0.50 max.
Water content	500 ppm max.
Sulfur content	0.05 % max.
Total glycerine content	0.24 % max.
Total contaminants	24 ppm max.
Total alkali metals	5 ppm max.
Oxidative stability	6 hours
Iodine value	120

Table 4-1 : Requirements that must be met to fulfill the EN 14214 standard include minimum water, sulfur, glycerine, alkali metals and other contaminants while meeting property standards such as oxidative stability, iodine value, acid value and cetane number.[145]

As shown in the table, many requirements must be met in order to practically utilize biodiesel as a fuel in Europe, the United States and other countries. Biodiesel manufacturers, in general, require that some type of quality control testing be performed in order meet biodiesel accepted standards. Therefore, biodiesel testing facilities should have specific testing protocols that quickly determine the extent of contaminants, other substances and the values of required properties. These properties consist of oxidative stability, iodine value, acid value and cetane number. Minimum contaminant requirements are also standardized. They include certain minimum concentrations of water, sulfur, glycerine, alkali metals and other contaminants (i.e., soaps) as shown in the table.[145]

Biodiesel has certain superior physical and chemical properties over petrodiesel. For these reasons its use in the overall fuel infrastructure is desirable. Biodiesel has little to no sulfur compounds or aromatics in the fuel whereas regular diesel fuel contains a large percentage of aromatics in them. In addition, it requires an expensive and additional hydrodesulfurization process in order to lower the sulfur content to 15 ppm (known as ultra low sulfur diesel [ULSD]) in petrodiesel. Other superior qualities biodiesel has over petrodiesel include better lubricity and improved exhaust emissions. It is also non-toxic and biodegradable. Biodiesel degrades at least 4 times faster than petrodiesel when exposed to the environment. It has been shown that the blending of biodiesel with petrodiesel fuel is helpful towards improving vehicle pollutant emissions. Blending biodiesel at percentages of around 20 % (B20) into regular diesel reduces particulate matter (PM), hydrocarbons and carbon monoxide when compared to emission pollutant values recorded for just straight petrodiesel.[146,147]

Average values of HFFR (lubricity wear scar) upon comparison of petrodiesel to biodiesel

Lubricity wear scar – HFFR	
Biodiesel	190 microns
Petrodiesel	550 microns

Figure 4-2 : The lubricity of a fuel is measured by a wear scar test that basically exerts pressure on a metal ball that leaves an impression on a metal surface.[157] Petrodiesel has a much lower lubricity than biodiesel as shown by the comparison of the wear scar size between biodiesel and petrodiesel.[149]

Currently biodiesel is primarily mixed into petrodiesel in order to improve its lubricity. The newer petrodiesel (ultra low sulfur diesel [ULSD]) has the disadvantage of suffering from low lubricity. The lubricity value of a fuel is important in terms of testing potential wear and breakdown to certain vehicle parts as measured by the lubricity wear scar or HFFR. The test determines the HFFR value by exerting a certain force on a metal ball immersed in the diesel fuel and then measuring the size of the wear scar etched by the ball on a metal surface as shown in Figure 4-2.[148]

Before ULSD was created, petrodiesel did not have issues with vehicle wear due to low lubricity. However, as soon as ULSD was manufactured in Europe two decades ago, they started to notice vehicle based problems. For example, Sweden in 1991 switched to a very low sulfur and aromatic diesel fuel (much like ULSD in the US) as required by government mandate. However, the use of this fuel caused eventual breakdowns in many vehicles due to the low lubricity of the newer diesel fuel.[150] The current newer US ULSD diesel at a minimum requires special fuel additives in order to increase the lubricity of the diesel fuel.[150] Therefore, the blending of biodiesel with petrodiesel fuel has brought about beneficial qualities that increase the overall lubricity of the fuel, significanty increasing its lubricity to accepted wear scar (HFFR) values.[149]

This section has shown that biodiesel has many advantages when compared to petrodiesel. In addition, biodiesel has the advantage in that the whole production of it, including the manufacture of the required alcohols, could all be done with renewable resources. Recall that biodiesel production requires the addition of an alcohol during the esterification based manufacturing process. The alcohols chosen are usually methanol or ethanol. These alcohols can be made from biomass using thermochemical gasification. The use of ethanol offers even more choices related to the types of production methods involving renewables.

4.3 Increasing biodiesel production could help to spawn a glycerol based manufacturing industry

A major stumbling block that has been preventing increased production of biodiesel is being able to find large enough markets for the waste glycerine. Being able to sell the glycerine is an important means of recovering the cost of biodiesel production. The manufacturing process yields around 10 percent waste glycerin as a byproduct [1 pound glycerine per gallon of biodiesel]. In other terms, every 10 gallons of biodiesel makes one gallon of waste glycerine when taking into account its density (1.26 g/mL - glycerine). At high biodiesel production levels there is usually an excess amount of glycerine made. Oftentimes it is difficult to find enough uses for this waste glycerine shortly after it has been produced. As shown in Table 4-2, a 5 % biodiesel blend in regular diesel fuel, just in the US, would create a surplus of glycerine that would exceed even its current world demand.

Just the production of B5 would produce over 300 million gallons of waste glycerine while B20 would produce around 1.2 billion gallons of it. The current estimated worldwide demand for glycerol itself is only 50 million gallons [ie 500 million pounds][151] or around 6 times less the amount that would be made from B5 production in the US. An oversupply of the waste glycerine and its resulting inability to properly appropriate the material to other market sectors has caused its economic value to decrease during the initial boom of biodiesel production during 2007. Comparing the price of crude glycerol sold per pound during that year (2007), the price of crude glycerine dramatically decreased from 25 cents per pound to 5 cents per pound.[152]

Supply and demand of glycerol as determined by possible increased production of biodiesel in the United States per year

Fuel Grade	Waste glycerine produced
B5	300 million gallons
B20	1.2 billion gallons
Current World Demand	50 million gallons

Table 4-2 : A 5 % or 20 % biodiesel blend in regular diesel (called B5 or B20) would create an additional 300 million to 1.2 billion gallons of waste glycerin, which greatly exceeds the current world demand of 50 million gallons used per year.[151]

What if a significant market for waste glycerine was developed? This is already happening. Waste glycerine from biodiesel has already found some market niche areas in industries such as pharmaceutical, food, cosmetics and soap manufacturing. Glycerol may also be utilized as a starting ingredient for products such as printing inks, perfume, surfactants, paints, coatings, paper, environmental products, synthetic foams, lubricants, food additives and plastics.[153] Certainly other newly applied uses for glycerol would help to increase its value as a potential commodity chemical, which in turn would assist with the increased economic based development of biodiesel. Many scientists agree that glycerol as a starting chemical for product development has around a thousand or more uses.[152] Indeed, many of these uses have not yet been developed, although they are mentioned frequently in the scientific literature.

According to government studies, glycerol is considered to be one of the top twenty renewable based chemicals useful towards synthesizing petrochemical substitute compounds.[163] Figure 4-3 below demonstrates how glycerol forms one of a number of other chemical building blocks through different reaction synthesis schemes and also points out what some of these compounds are utilized for.[153,155,156] Through such methods of chemical synthesis, glycerol has the potential to create a diverse subset of compounds useful in manufacturing other consumer or industrial products.

Chemical building blocks used in consumer & industrial products obtained through synthetic reactions with glycerol

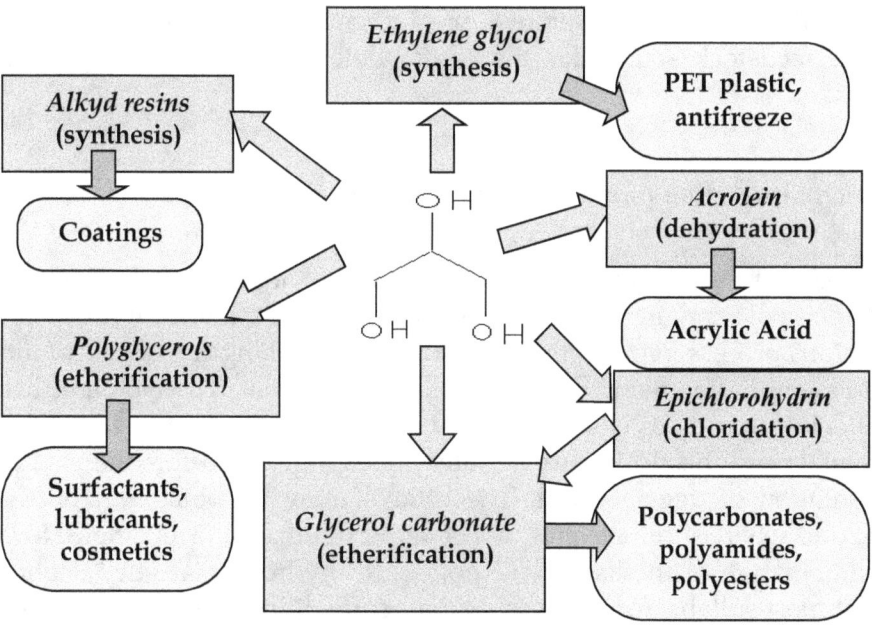

Figure 4-3 : A large number of chemical byproducts can be made from a subset of chemical building blocks originally made from glycerol through synthesis schemes. Many of these chemicals are utilized for plastics or industrial products like surfactants, lubricants and coatings.[153,155,156]

For example, the chemical building blocks of glycerol carbonate, acrylic acid, alkyd resins and ethylene glycol can all produce other types of plastics. All of these chemical building blocks are originally derived from glycerol. Glycerol carbonate can produce plastic materials that consist of polycarbonates, polyamides and polyesters. The manufacture of ethylene glycol can assist towards making the plastic polyethylene terephthlate (PET). In addition, other compounds such as polyglycerols have the potential to make certain types of surfactants, lubricants and cosmetic formulations.

Therefore, it should be apparent that glycerol has great potential towards making chemicals utilized for many product based formulations. Since these basic chemical building blocks only require one to two synthesis steps that start with glycerol, it would be practical to locate basic chemical synthesis factories near the source of the waste glycerine, i.e., close to biodiesel manufacturing refineries.

In addition, the practical use of glycerol as a carbon source for microbial based fermentations should not be undervalued. Glycerol, like glucose, could be provided as a major feedstock source for the cultivation of many microbial, yeast and algae species. For example, one renewable diesel manufacturing company in California claims that glycerol is one of the feedstock sources they use to cultivate algae for biofuel production.[157] Also, glycerol provides a possible feedstock source for the carboxylate counter-current fermentation process.

4.4 Batch description of manufacturing biodiesel from vegetable oils or fats

Biodiesel oftentimes takes place as a batch process much like baking a batch of cookies where several dozen would be made at one time. A continuous process, on the other hand, refers to making cookies constantly, as in an assembly line using the appropriate equipment. The biodiesel batched manufacturing process is shown below in Figure 4-4. In the first step [1] both the catalyst and the alcohol are initially mixed together before they are added to a solution of vegetable oil. The catalyst may be potassium or sodium hydroxide whereas the alcohol can be either methanol or ethanol. However, the combination of methanol and potassium hydroxide are the alcohol and catalyst most suitable for biodiesel manufacture utilizing a batch process.

After the alcohol and catalyst have been premixed, they are added to a batch reactor that contains the vegetable oil [2]. The batch reactor should be outfitted to provide the appropriate temperature and mixing conditions. An important consideration is the ratio of alcohol mixed with vegetable oil. For a normal batch process, the oil is at least 4-5 times the amount of alcohol inputed into the reactor. During step two [2] these components are mixed together at a temperature above 60 degrees Celsius for around 30 – 60 minutes.[145,158] After this initial reaction takes place, the resulting biodiesel products of methyl esters and glycerine are allowed to settle and separate from one another.

Diagram of the batch process involved in producing biodiesel from vegetable oils or fats

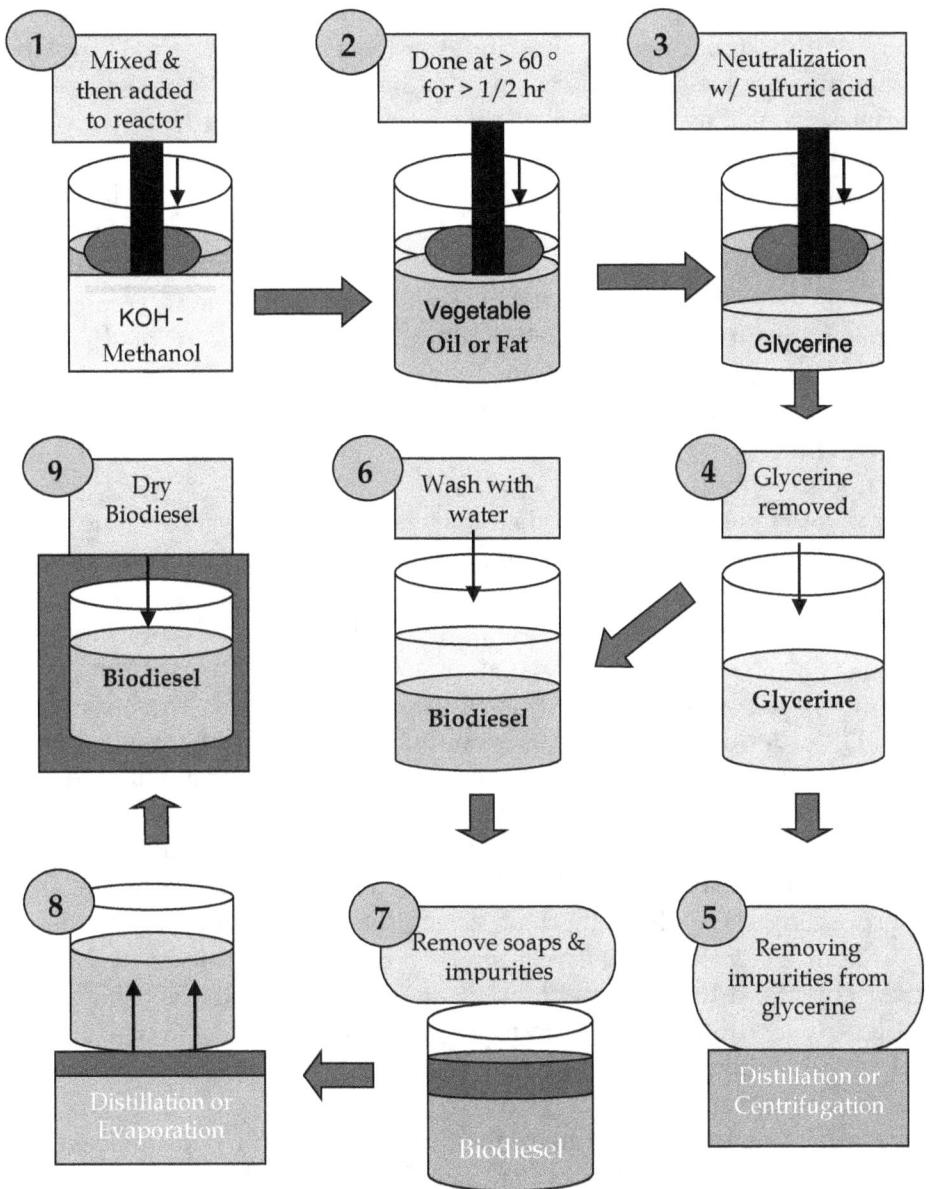

Figure 4-4 : This diagram depicts the necessary steps involved in producing biodiesel as a batch process, which is the method most often utilized. Steps are numbered in the diagram not necessarily in required sequence.

During the third step [3] the acidity of the mixture is tested to see if it requires neutralization. The neutralization process involves adding sulfuric acid to the biodiesel and glycerine mixture. The acidity test involves a titration method that effectively measures the concentration of the free fatty acids contained in the mixture after the biodiesel reaction has gone to completion. This neutralization step reduces and gathers the free fatty acids in the biodiesel so that they go into the glycerine fraction. This helps to prevent soapy compounds from forming during the washing step. In step four [4], the glycerine gets separated from the biodiesel by simply draining it from the bottom of the reactor. The separated glycerine fraction will also contain some impurities as well.

In step five [5] the impurities contained in glycerine are removed by performing distillation or centrifugation. The purification of the waste glycerine happens to be one of the more important steps that take place during biodiesel production. Helping to purify the waste glycerine above 80 % purity will prepare it for future sale as a basic commodity utilized for building other important products.[159]

Next [6], a good wash with distilled water helps to remove many of the contaminants in the final biodiesel mixture, including the soaps that buildup from the washing process. Washing is oftentimes executed several times to assist with better purity. In step seven [7], the biodiesel and water mixture go through another separation step that removes soaps and other impurities. Afterwards, the biodiesel requires additional distillation/evaporation step(s) [8] & [9] in order to remove excess amounts of water and methanol from the biodiesel. Once these steps have been completed the biodiesel will be ready to be used as a fuel source or it can be stored in a safe location.

It should be mentioned that advancements in technological development are now making it easier to produce biodiesel as a batch process. For example, new processing steps allow the biodiesel to be cleaned without the need to wash it with water. Water washing tends to be more labor intensive and less efficient compared to newer separation and cleaning methods. These newer methods involve filtering the biodiesel with materials like wood chips. In addition, carbonaceous material may be further utilized to remove other contaminants as well.

In today's world the manufacture of biodiesel mainly takes place in large volume scale refineries rather than in facilities with small scale batch production equipment. Oftentimes a commercial plant will still manufacture biodiesel as a semi-batch process by implementing more sophisticated equipment such as larger and more robust reactors, settlings tanks and pumps.[145] In addition these plants also contain larger separation equipment such as distillation columns and centrifuges.

4.5 Improvements in biodiesel processing have allowed for more continuous based production that yields higher volumes of biodiesel manufactured per year

Developments in technology within the last several decades have allowed biodiesel to be manufactured at higher volumes in refineries at the scale of millions of gallons per year. This section outlines several examples of these technologies. The newer technologies offer three main advantages over conventional biodiesel production. One main advantage involves efficient alcohol use during production. In these refinery settings the alcohol becomes recycled and utilized over and over again throughout the manufacturing process. The alcohol gets removed through distillation or evaporation and then becomes incorporated back into the original reactor that performs the esterfication process. This takes place as a continuous cycle, as shown in Figure 4-5. Most of these refineries use a large amount of alcohol in the circuit process. Opposite to batch mode production, the alcohol solvent reacts with the oil feedstock at a high mixture ratio with the alcohol being 5-10 times higher in volume than the oil.

A second advantage concerns manufacturing the biodiesel without the use of the water normally implemented in the washing stage of batch mode production. The absence of water input for refinery processes allows for more effective distillation and recycling of the alcohol (methanol).

The main improvement in production technology, however, has been the ability to directly convert free fatty acids into methyl esters as part of a multi-stage process, as demonstrated in Figure 4-5. A two step catalytic based conversion method converts both the free fatty acids (FFAs) and triacylglyeride (TAGs) material into methyl esters. Oftentimes this process incorporates two different types of catalysts being added to each of the respective stages, as shown in the figure.

One example of this technology includes a multi-stage two solvent system developed by the company BIOX. It incorporates both methanol and another non-toxic solvent known as tetrahydrofuran (THF).[160] This production method also implements the same two-step catalytic conversion process, just described above, that converts the free fatty acids (FFAs) and triacylglyceride (TAGs) material into methy esters. Next, the biodiesel/glycerine/solvent mixture then goes through an evaporator where the two solvents (methanol & THF) are separated from the mixture. These solvents then get recycled back into the initial mixing and reaction process through the feedback loop shown in Figure 4-5. Finally, the remaining biodiesel/glycerine then is transferred to a separator where the glycerine is removed from the biodiesel.

This process has several advantages when compared to the normal biodiesel batch production technology.

- The reaction rates of converting the oil into esters is several times faster.[161]
- The separation of glycerol from the biodiesel also takes place much faster and is more effective at separating the contaminants into the glycerol phase.[162]
- The method utilizes a solvent (THF) with methanol that is non-toxic and can be made from natural biomass sources.

Two solvent or methanol semi-continuous biodiesel production system

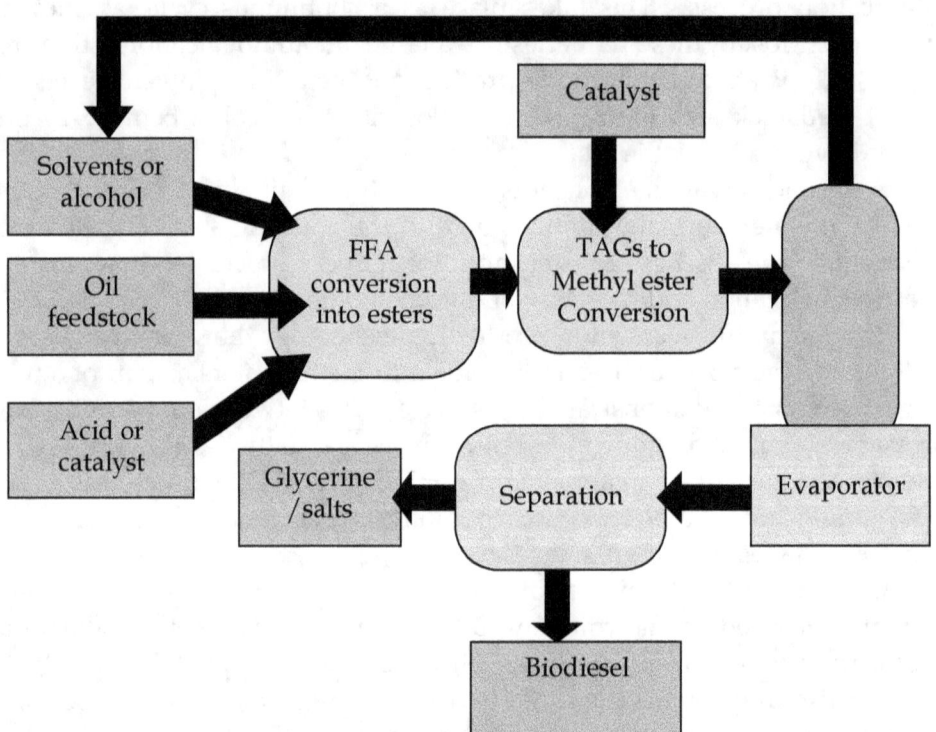

Figure 4-5 : A methanol or two solvent biodiesel manufacturing process executes a two-step process while recycling the solvents and also giving more effective separation.[160]

A similar semi-continuous system developed by the company Pacific Biodiesel Inc. also executes a multi stage mixing and reaction process to convert a feedstock containing a high percentage of free fatty acids into biodiesel.[159] The process incorporates a similar two stage catalytic conversion system that also efficiently recovers and puts methanol back into the production stream via distillation. In addition, the company implements dry refining and filtration stages that effectively remove the impurities from the biodiesel. This production system also offers several advantages over regular batch mode biodiesel manufacture:

- It includes the redesign of a production facility to produce from 0.5 to 10 million gallons of biodiesel per year. This takes place by retrofitting an existing facility with their own proprietized plant equipment.[159]
- This technology accepts a wide range of feedstock sources including yellow or brown grease, tallow or canola oil. In fact it allows a feedstock source that contains up to 50 % free fatty acids to be utilized.[159] This means that one can incorporate brown grease into the overall production process.

Ultrasonic shearing biodiesel production technology

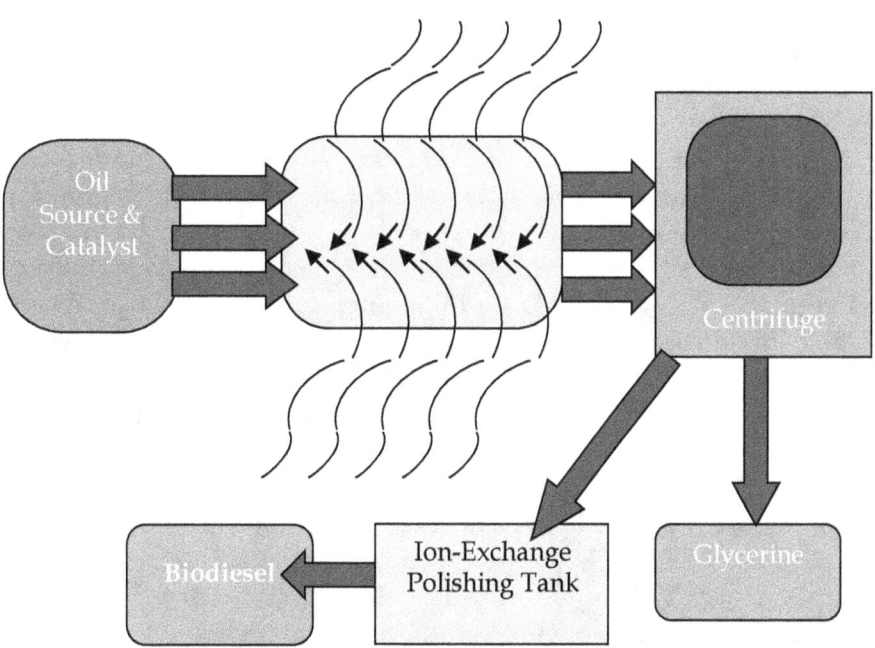

Figure 4-5 : The ultrasonic method inputs sound waves that aggresively react with the oil and catalyst to produce biodiesel within a short period of time. This method overall also claims to be continuous in method.[163,164]

Another revolutionary technology of recent development includes a sonication production platform that converts the vegetable oil directly into biodiesel when coupled with a catalyst. The production method is continuous due to the very fast reaction rates that take place. The oil and catalyst react very rapidly with the sound waves from the sonication process, which has been claimed to happen within a very short period of time.[163,164] This method also requires a minimal amount of energy for processing when compared to conventional biodiesel manufacturing. As shown in Figure 4-6, a typical refinery setup would include a sonication based reactor, centrifuge for glycerine separation and ion-exchange equipment for the removal of impurities in the crude biodiesel mixture.[163] The majority of the companies involved with this technology are still in the pilot planning stages of production. One company named IncBio also has the ability to retrofit and upgrade the older biodiesel facilities with the more advanced sonication based equipment.[164]

4.6 Biodiesel can be made through biological based enzymatic production systems

Biodiesel can also be produced through biological means where enzymes carry out the necessary reactions required to make methyl esters from oil sources. These types of enzymes are called lipases. Lipases are known both as whole microbes that contain 'lipase' enzymes or are the enzyme entities themselves extracted from these types of microbes. Either definition applies towards biodiesel production.

Biodiesel production performed by lipases has been done on oil sources such as vegetable oil, waste cooking oil, animal fats and other processed oil wastes.[165] Other processed oil wastes include the residuals leftover after processing vegetable oil into cooking oil. The processes utilized to make the cooking oil are oil seed extraction, degumming, bleaching, neutralization and deoderization.

The lipase based method of converting vegetable oil into alkyl esters is outlined in Figure 4-7. The lipase enzyme takes a fatty acid and converts it into an alkyl ester. These lipases perform the same function that chemical catalysts such as hydroxides do. In other words they facilitate the esterification reaction that converts a fatty acid into an alkyl ester. Similar to chemical catalysts, lipases also require the presence of an alcohol such as methanol so that the esterification reaction takes place. Lipases that are utilized towards biodiesel production offer an additional advantage. The lipases have the ability to simultaneously convert both the FFAs and TAGs into methyl esters. In essence this removes the need to have a two step catalytic conversion process or other FFA removal step.

In addition, the form in which the lipases are provided to the reactor vessel tends to be an important consideration for this method of biodiesel production. Lipases can be established in the reactor vessels as the free form or an immobilized form within the reactor. However, research has shown that immobilized systems may be preferential in application for economic and practical reasons. For example, lipases embedded in membranes, gels or resins can be used several times, are biologically stable and are more easily translatable towards higher scale industrial based production with implications of processing algal oils as well.[166] Also, literature sources claim that lower costs are involved with the growth and application of immobilized lipases for biodiesel production versus the free lipase form.[167] In addition, the immobilized lipases can be reused several times and do not significantly accumulate in the oil source. This makes it easier to separate out the lipases afterwards and then reutilize them.

Production of alkyl esters from fatty acids through biocatalytic conversion involving lipases and an alcohol

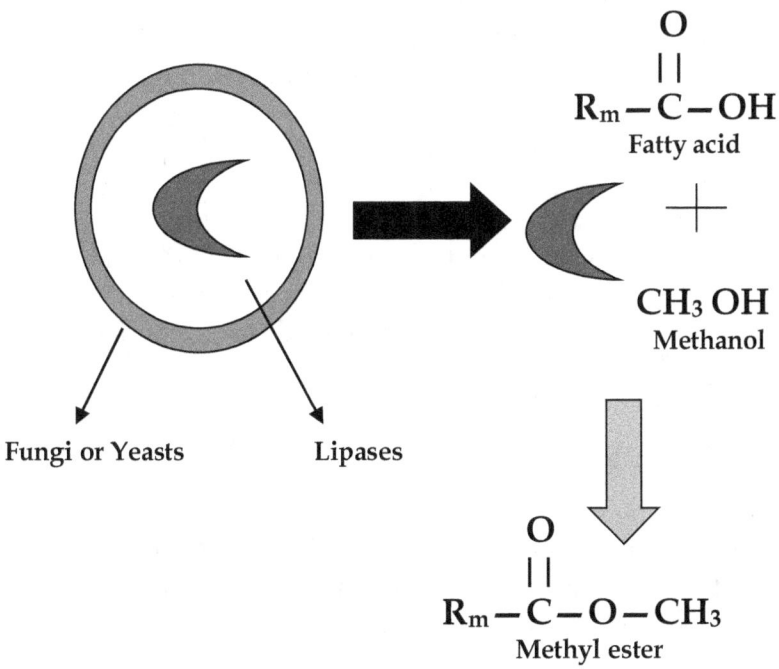

Figure 4-7 : An alternate production method that is used to make alkyl esters involves mixing vegetable oil or wastes with enzymes called lipases. Lipases directly utilize an enzymatic reaction that converts a fatty acid into an alkyl ester. It is also required to provide an alcohol such as methanol when utilizing the lipase biodiesel production method.

A readily available and ample supply of lipases assists towards making the biodiesel effectively. However, similar to other enzymes systems, the cost of purchasing lipases can be expensive, thus making this production process somewhat prohibitive. Therefore, ways to make these enzymes in a more economical manner needs to be discovered. This could involve finding accessible oil waste sources that are attached to the biodiesel production process itself. One potential source involves the utilization of the leftover seedcake from the oil extraction process. The seedcake itself still contains at least 10 % or more of the original seed oil which would do well to serve as a potential oil source for lipase production. Researchers have suggested that the cultivation of lipases from oilseed cake would create a monetarily cheap supply of them.[168] In addition, as an important sidenote, lipases have been known to remove toxic compounds contained in non-edible oilseed cakes.[168]

4.7 B100 has disadvantages that can be corrected with the use of fuel additives

In addition to important advantages, biodiesel has its specific disadvantages for use as a vehicle fuel. For example, since biodiesel generally has a higher viscosity value than petrodiesel, it causes higher amounts of unburned hydrocarbons to accumulate within the engine cylinder walls and pistons. The unburned hydrocarbons can even get past the pistons and then become mixed into the lubrication system itself causing it to sludge over (dirtying, thickening & clumping of the oil) in time as shown in Figure 4-8. Biodiesel also has the disadvantage in that it cannot be stored for long periods of time due to its chemical modification from aging and oxidation stability issues. On the average biodiesel has a recommended storage period of around 6 months up to a year.[148]

If the biodiesel becomes destabilized due to aging or oxidation it can cause gumming and sediments to form in the fuel, which then adversely affects the fuel lines and fuel pump. In addition, prolonged aging of the biodiesel causes an increased solvent effect that can dissolve sediments into the bottom of the fuel tank. Sediments can also form in B100 if the biodiesel has a high cold filter plugging point (CFPP) temperature, meaning that at very cold temperatures the biodiesel also tends to clump up and block the fuel lines and pump.

The corrosive nature of biodiesel also causes compatibility problems with certain plastics or rubbers found in the engine as parts such as hoses, gaskets and seals. As shown in Figure 4-8, incompatible materials associated with B100 include plastics such as polyvinyl, tygon and polypropylene while plastics like Teflon, Viton and nylon are compatible materials for 100 % biodiesel.[148]

The specific disadvantages of B100

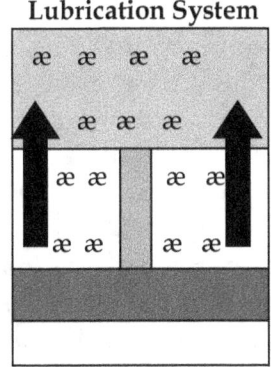

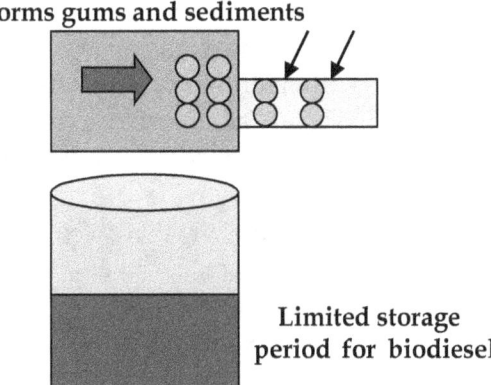

Residual unburned hydrocarbons accumulate in engine & lubrication system with biodiesel

Sediments and gums form in fuel lines and fuel filters while storage shelf life is limited to 6 months due to oxidative stability

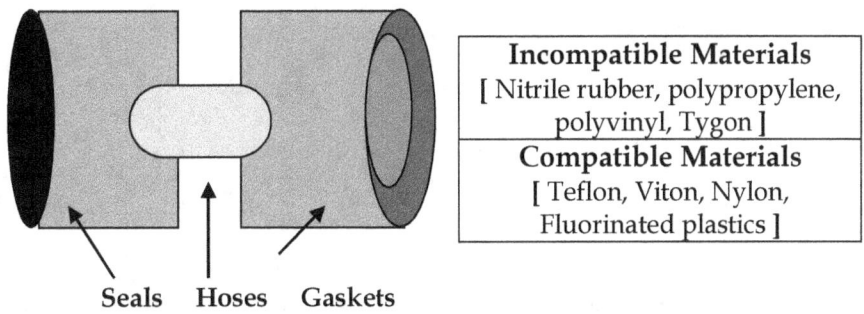

1) B100 compatibility/incompatibility with plastic/rubber materials
2) B100 also degrades or leaks through seals, gaskets & hoses

Figure 4-8 : Biodiesel at 100 % (B100) has specific disadvantages when it comes to its use in vehicles. Biodiesel combustion accumulates hydrocarbons within the vehicle, it can be incompatible with specific plastics/rubbers even to the point of causes leaks through seals, hoses, etc. Biodiesels oxidative stability/age can cause the fuel to change chemically and therefore, give a limited shelf storage period

Therefore, the incorporation of 100 % biodiesel (B100) in most diesel engines is not recommended due to vehicle wear issues related to the stability, aging, cold temperature behavior, viscosity and solvency of the fuel itself discussed previously. However, measures such as producing biodiesel with differing feedstock sources and placing fuel additives into biodiesel can ameliorate these adverse vehicle wear issues. Specifically, fuel additives have been shown to improve storage and chemical stability, cold temperature behavior, viscosity and decreased chemical incompatibility, making it practical for vehicle operation.

As a point of encouragement, B100 is attractive for use in smaller market sectors such as the operation of farm machinery equipment. During the last decade farm machinery has been outfitted to operate on B100. This works out very well with landowners like farmers since they, potentially, could also produce their own biodiesel utilizing the natural resources from their land. It has been shown that tractors have been able to run on B100 at a university farm for several years without showing adverse performance issues.[169] Other types of farm equipment can regularly operate on B100. This includes combines, windrowers, sprayers, etc. made by companies like John Deere, New Holland and Case IH.[170,171,172] Some of these manufacturers state that their diesel engines can operate on biodiesel mixtures from B20 to B100 but require that the engines do not contain an exhaust filter.[171]

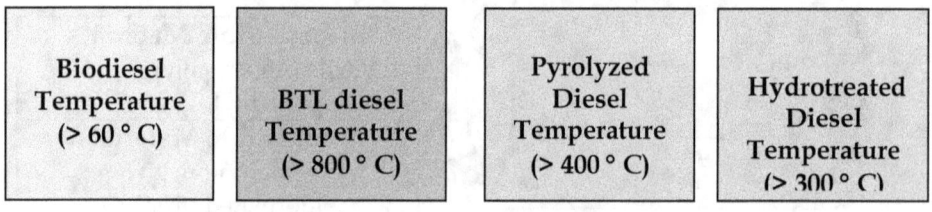

Figure 4-9 : Comparisons between the processing temperatures required for the various renewable diesel processing technologies such as biodiesel, BTL, pyrolyzed and hydrotreated diesels.[158,175,176,177]

Biofuel operation with important farm equipment like tractors has important implications as a market niche since well over 100,000 of them sold in the United States every year.[173] Due to need for cleaner emissions, biodiesel may also find a future application with other types of larger stationary diesel engines. The EPA has become more strict with the emission based standards starting around 2014-2015 in states such as California.[174]

Also, the biodiesel manufacturing process has an energy based advantage over other types of diesel manufacture. Biodiesel production is considered to be a very energy efficient process due to its low thermal energy requirement. The processing temperature associated with biodiesel is much lower when compared to all other forms of renewable diesel manufacture as shown in Figure 4-9. Biodiesel manufacture requires processing temperatures usually ranging from 60 to 100 ° C while other renewable diesel processing technologies operate their reactors at temperatures of at least 300 °C or higher. Hydrotreated and pyrolysis based diesel operating temperatures start at around 300 or 400 °C and go higher for each respectively.[175,177] Biomass to Liquid (BTL) gasification processing temperatures are even higher usually around 800 °C or higher.[176] Processing temperatures for petrodiesel manufacture are also very high in range. Therefore, due to its lower processing temperature, biodiesel manufacture has a very high net energy value (NEV) when compared to other types of petroleum or biofuel production models. Biodiesel yields a net energy gain of at least 3.2 times the energy required to produce it while petroleum based manufacture for vehicle fuels actually gives a negative return in energy value (around 20 %).[178]

4.8 The performance properties of biodiesel are determined by the type of oilseed or fat source one utilizes

Biodiesel can be made from a large array of biomass sources that include vegetable oils, animal fats and algae. Table 4-3 gives a comparison between the specific properties, performance and oil yield per acre of some common vegetable oil and animal fat sources that are cultivated and utilized for biodiesel production. As shown in the table, the oil yield per acre varies greatly between these oilseed sources. For example, corn oil averages 18 gallons per acre while palm oil is known to produce around 600 gallons per acre. Soybeans as a crop give a low yield per acre at around 50 gallons per acre.

Oilseed based trees seem to give higher yields of oil per acre in general. This applies with both palm dates and jatropha. Both trees prefer to grow in warm climates. In fact, jatropha usually has to be cultivated south of the United States because it must grow in non-frost conditions. There are other oilseed sources that grow in the southwestern deserts of the US and also give high oil yields per acre. These include plants such as jojoba (260 gallons/acre) and castor beans (170 gallons/acre). In addition, many municipal waste sources can be recovered in appreciable amounts in most US cities. These include waste cooking oil, waste grease and animal fats. Their conversion into biodiesel should be able to produce well over 2 billion gallons of it per year as shown in the table.

The effectiveness of a biodiesel fuel derived from a specific oilseed source depends upon its performance related properties. Some of these properties are **cloud point temperature, viscosity** and **cetane number**. Table 4-3 outlines the measurable values for these properties specific to each type of oilseed or fat source. Probably the most important property that determines effective diesel engine performance is the cetane number. Cetane number measures the combustion quality of a diesel fuel during compression ignition. Other measurable values such as ignition delay and combustion duration help to establish a fuel's cetane number. The higher the cetane number the better the fuel will perform in a diesel engine.

Most of the oilseed sources have very good cetane numbers indicating that, in general, the ignition quality of the biodiesel made from them is excellent. As shown in the table, cetane number varies somewhat between oilseed sources. Palm oil, corn oil and jojoba oil have higher cetane numbers in the 60's. Other oilseeds usually have average cetane numbers ranging in the high 40's to the 50's. As a general rule, any fuel source that has a cetane number above 40 tends to perform very well in diesel engines. Petrodiesel in general tends to just have average cetane values. Therefore, oilseed based biodiesel usually has a better cetane value and resulting combustion performance when compared to petrodiesel.

Comparison of the important properties, yield per acre and cetane number between various types of oilseeds

Vegetable oil source	Yield per acre	Cetane number	Viscosity	Cloud point
Palm dates	610 gallons / acre	62	3.90 mm^2 / s	13 °C
Soybeans	46 gallons / acre	46	4.10 mm^2 / s	0 °C
Corn	18 gallons / acre	60	4.19 mm^2 / s	-3 °C
Jatropha	260 gallons / acre	52	4.50 mm^2 / s	2 °C
Rapeseed	100 gallons / acre	47	4.60 mm^2 / s	-3 °C
Waste cooking oil	**~ 2 billion gall. per year	44	4.49 mm^2 / s	8 °C
Animal tallow	**> 500 million gall. per year	58	4.10 mm^2 / s	17 °C
Castor beans	170 gallons / acre	49	15.17 mm^2 / s	-14 °C
Jojoba	120 gallons / acre	69	6.67 mm^2 / s	-13 °C

Table 4-3 : At least 9 different oilseeds, waste cooking oil or animal fats are compared with one another as far as yield per acre, cetane number, viscosity and cloud point temperature. The important factors are yield per acre and acid number as these determine ease and amount of biodiesel that is produced.[179,180,181,182,183,184,185]

** Quantities are applicable just in the United States

Another property that determines beneficial low temperature operation of the biodiesel in an engine is cloud point temperature. Cloud point temperature can be measured as the lowest temperature at which the compounds within the biodiesel gel or thicken, causing the fuel to stop flowing normally in the engine. The lower the cloud point temperature the better the fuel responds to lower cold weather temperatures. Desert oilseed sources such as castor beans and jojoba have excellent cloud point temperatures as shown in the table. On the other hand, oil sources that have higher cloud point temperatures do not respond beneficially to cold weather conditions such as freezing temperatures. Therefore, cloud point temperature values greater than $0\ °C$ for some types of oilseeds involve the risk that the fuel will gel or thicken. This type of behavior is more evident with sources such as waste cooking oil, animal tallow and palm dates.

Another important property relating to biodiesel efficacy is its viscosity. Viscosity values help to determine the combustion completeness of a fuel source. The trend in general shows that low viscosity values equate to better overall completeness of combustion. More complete combustion correlates to less unburned hydrocarbons that become part of the exhaust emissions or that get into the lubrication system. Viscosity also measures the flow behavior of a fuel. Lower viscosity values usually indicate that the flow properties of the fuel behave more beneficially. In other words, the fuel tends to flow better at lower viscosity values.

It should be pointed out that the viscosities of these oil sources are based upon the individualized methyl esters rather than the fatty acids contained in the unprocessed triacylglycerides. Therefore, fuel researchers tend to utilize methyl ester viscosities rather than the viscosity from the raw oil source itself. Interestingly, the oil sources themselves are usually several times higher in viscosity value when compared to the combination of their respective methyl esters. The viscosities of most of the methyl esters obtained from various oilseed sources are similar in value, usually ranging from $4 - 5\ mm^2/s$. Only castor bean and jojoba methyl esters have much higher viscosities (greater than $6\ mm^2/s$) as is shown in the table. Regardless, the viscosities of these particular methyl esters are greater in value than the average viscosities seen in both petrodiesel and other forms of renewable diesel. Therefore, biodiesel in general could use lower viscosity fuel additive compounds in order to improve its viscosity value.

The majority of biodiesel contains normal performance property values. Therefore, it usually behaves as it should. This is typical of biodiesel from oilseed sources such as corn, soybeans, jatropha and rapeseed. However, other types of biodiesel can have a diverse set of fuel performance properties. Certain types contain inversely related viscosity values and cloud point temperatures. For example, the desert oilseeds of castor beans and jojoba shown in Table 4-3 have high viscosities but yet give lower cloud points (less than -10 ° C). On the other hand palm oil has a very low viscosity value but high cloud point temperature. In other words these types of biofuels either have a beneficial cloud point temperature or viscosity value. Therefore, the fuel is not desirable in both areas.

One then may ask why the combinations of properties in certain oilseed sources behave so differently from one another. The answer has to do with the effect that the individual methyl esters have on each one of these properties. These property values tend to fluctuate a bit upon comparison between the methyl esters themselves. The next section discusses how an oilseed's overall property values can be estimated based upon their specific methyl ester's property values. It suffices to say that these values can be ascertained from the percent composition of each fatty acid contained in the specific oilseed source.

4.9 Weight averaging the performance properties of each respective methyl ester gives a better estimation of biodiesel fuel performance from a given oil source.

There is a method that accurately estimates the values of viscosity, cetane number and cloud point temperature of biodiesel for an oilseed type based upon its specific fatty acid (or methyl ester) composition. This can be accomplished by weight averaging the performance property values that correspond to the specific methyl esters contained in a given oil source. One obtains the weight average by multiplying the percentage amount of a specific methyl ester with its corresponding property value and then summing up the results from each type of methyl ester. One then divides the result by the number of methyl esters added together.

Table 4-4 provides the individual cetane numbers, kinematic viscosity and melting points of five methyl esters commonly found in vegetable oils.[186] By doing a quick analysis before calculation, biodiesel performance properties can be predicted by examining the basic differences between certain methyl esters. For example, methyl palmitate has a very high cetane contribution yet has too high of a melting temperature, a quality that gives the fuel a poor low temperature behavior. An oil source could also contain the ester methyl linoleate that has a lower cetane value yet gives an excellent low temperature melting point. So, it is important to note that a given oil source behaves closer to the higher concentrated methyl ester that is contained in it. Thus, a weight average of the top three or so fatty acid (methyl ester) property values gives an accurate estimation of an oil sources overall performance behavior.

The performance properties of cetane number, kinematic viscosity and melting point of certain methyl esters

Type of methyl ester	Cetane number	Kinematic Viscosity	Melting point
Methyl palmitate	86	4.38 mm²/s	28.5 °C
Methyl oleate	59	4.51 mm²/s	-20.2 °C
Methyl linoleate	40	3.65 mm²/s	-43.1 °C
Methyl linolenate	23	3.14 mm²/s	-55 °C
Methyl erucate	74	7.33 mm²/s	-3 °C

Table 4-4 : The properties of cetane value, kinematic viscosity and melting point are given for five common methyl esters found in various vegetable oil sources utilized to produce biodiesel.[186]

Fatty acid distribution of a certain number of vegetable oils and animal fats

	Palmitic Acid	Linoleic Acid	Oleic Acid	Other FFA's [> 30%]
Canola oil	-----	19 %	64 %	----
Soybean oil	11 %	52 %	25 %	----
Jatropha oil	14 %	33 %	45 %	----
Lard	-----	15 %	50 %	33 %
Rapeseed	-----	15 %	22 %	> 45 %
Tallow	-----	-----	39 %	> 47 %
Palm oil	45 %	-----	39 %	-----
Waste vegetable oil	16 %	25 %	45 %	-----
Cottonseed	25 %	53 %	18 %	-----
Peanut oil	11 %	31 %	46 %	-----

Table 4-5 : The majority of vegetable oils and animal fats have a combination of linoleic acid and oleic acid with some also having a mixture of other unmentioned fatty acids or palmitic acid.[181,185,187,188,189]

Table 4-5 gives a percentage composition of the three most common fatty acids encountered in oilseed and animal fat sources. These fatty acids are oleic acid, linoleic acid and palmitic acid. Although most oilseed sources contain four or more fatty acids, the percentages of the other types are usually much lower in relative percentage content than the top three contained in an oilseed source. Oilseeds that mainly contain these three types of fatty acids appear to have median values of cetane number, viscosity and cloud point temperature. This includes oilseed sources such as soybeans, waste cooking oil, peanut oil, cottonseed oil, canola and jatropha. These types of oil sources also tend to contain a high percentage combination of oleic and linoleic acid. On the other hand sources such as animal fats have a large amount of other fatty acids (> 30 %) as shown on the far right column (other FFA's) in the table. This may in part explain why animal fats also contain higher cloud points than other vegetable oil sources.

The personal or industrial products that can be made from fatty acids or fatty alcohols

Health care products	Hand cream & lotions, shampoos, Shaving cream, hair products
Industrial products	Lubricants, coatings, plastics, elastomers, Surfactants, rubber, textiles, detergents

Table 4-6 : A large number of health care or industrial products contain specific fatty acids or fatty alcohols in them that can be obtained from various oil sources.

Value of fatty acids from oilseeds : The fatty acids in oilseeds are not only utilized for biodiesel production but are also a valuable source for some ingredients in other industrial products. Some of the health care and industrial products that have the special ingredients of fatty acids or fatty alcohols obtained from oilseed sources are listed in Table 4-6. These include shampoos, shaving cream, lubricants, coatings, surfactants and detergents, to name a few. Also, keep in mind that in the future algae oil sources could potentially provide a supply of fatty acids used in product formulations. The next section gives more details regarding the types of products formed from the fatty acids of certain types of desert oilseed sources.

4.10 Desert oilseeds are good sources for biodiesel production as well as ingredients in commercial products

As stated above, the oilseeds of jojoba, castor beans, jatropha, salicornia and palm dates can be grown in the desert regions of the United States and Mexico. This mainly includes the Sonoran and Mojave Desert regions. These desert oilseed plants can be classified according to whether they are edible or non-edible for human or animal consumption. Edible oilseeds that grow in desert climates include date palms and salicornia. Non-edible desert oilseeds include jatropha, jojoba and castor beans. Non-edible seedcakes have the possibility of being toxic to ones health if they are consumed. However, they may still be eligible for use in other products such as natural pesticides. Desert oilseeds can also serve as ingredients in consumer or industrial products in addition to making biodiesel regardless of whether they are edible or non-edible. Therefore, this section focuses on the relevant types of alternative products that are partially made by them.

Outline of palm date extraction byproducts and some of the material products that come from them

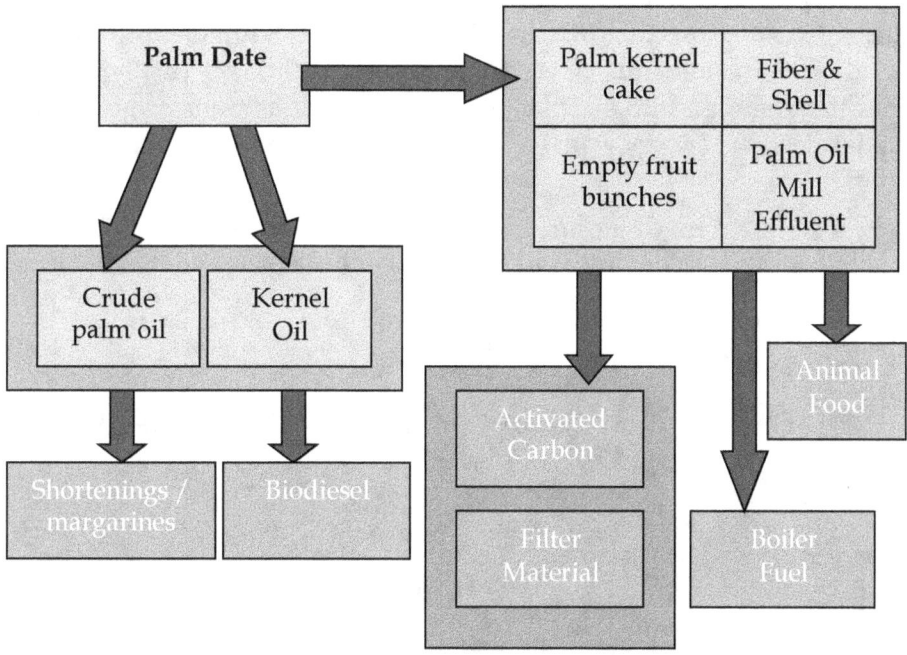

Figure 4-10 : The processing of the palm date itself creates many other products such as animal food, boiler fuel and activated carbon from the leftover byproducts of fiber, liquor, effluent and kernel cake. Palm oil itself is separated into two kinds of oil called crude palm oil or kernel oil, which can be converted to biodiesel or other food products.

One of the more interesting desert oilseed sources are date palms. Date palms potentially have a number of interesting environmental and competing product uses. The whole palm date produces several types of palm oils, practical byproducts and animal feed, as outlined in Figure 4-10. However, the waste byproducts themselves oftentimes have been disposed into the environment, causing a pollution problem. Nevertheless, byproducts such as palm kernel cake, empty fruit bunches, palm oil mill effluent and fiber & shell material have the potential to be recovered and converted into useful commercial products.[190,191] The kernel cake and fiber can further be processed into animal food. Waste products such as palm oil mill effluent can be sent to an Anaerobic Digester while fiber and shell material has the potential to be a source of boiler fuel.[190] Palm empty fruit bunches can be utilized to make activated carbon or fiber based filter material.[191,192]

There are two different types of palm oil obtained from the palm date. These include the kernel oil and the crude palm oil. Each type of oil has different properties due to their respective fatty acid content. As a general practice, the crude palm oil is best turned into biodiesel while the kernel oil becomes fashioned into food products such as shortenings and margarines. Another interesting product that has the potential to be made from the palm kernel oil is the manufacture of electrical insulating fluid.[193] This type of fluid becomes incorporated into electrical equipment such as transformers that are part of the electrical energy grid.

A methyl ester and fatty alcohol derived from a waxy ester in the jojoba oil

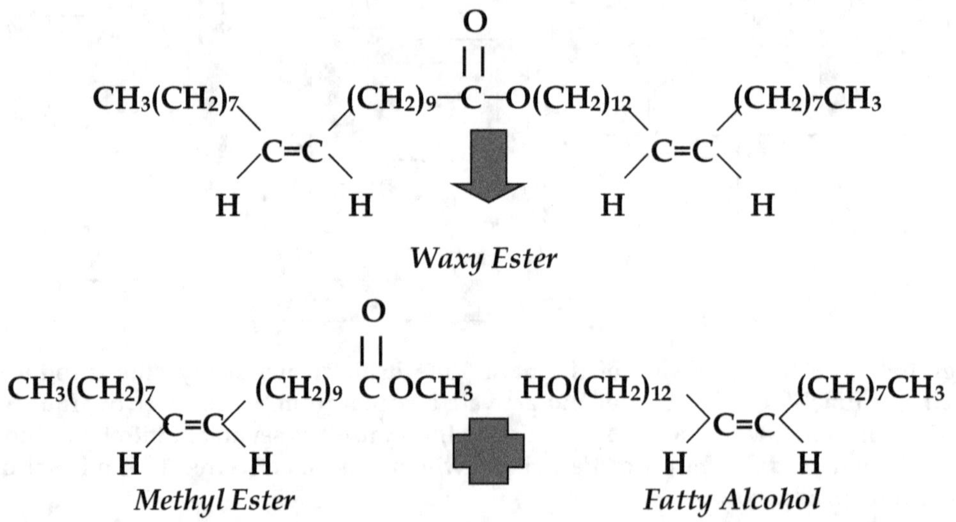

$$CH_3(CH_2)_7\underset{H}{\overset{}{\diagdown}}C=C\underset{H}{\overset{}{\diagup}}(CH_2)_9-\underset{}{\overset{O}{\underset{\|}{C}}}-O(CH_2)_{12}\underset{H}{\overset{}{\diagdown}}C=C\underset{H}{\overset{}{\diagup}}(CH_2)_7CH_3$$

Waxy Ester

$$CH_3(CH_2)_7\underset{H}{\overset{}{\diagdown}}C=C\underset{H}{\overset{}{\diagup}}(CH_2)_9\;\underset{}{\overset{O}{\underset{\|}{C}}}\;OCH_3 \quad + \quad HO(CH_2)_{12}\underset{H}{\overset{}{\diagdown}}C=C\underset{H}{\overset{}{\diagup}}(CH_2)_7CH_3$$

Methyl Ester   *Fatty Alcohol*

Figure 4-11 : Jojoba oil reacted with KOH and methanol create a methyl ester and a fatty alcohol

Biodiesel manufactured from jojoba is another ideal oilseed source that can be cultivated in fairly large yields per acre in the desert. The jojoba plant has been commonly grown in countries such as Israel, Mexico, United States, Chile, Australia, Peru and Egypt.[194] This oilseed source has the unique property of containing large molecular weight waxy esters that range in sizes from 22 carbons to 42 carbons. The waxy ester of C_{42} can make up greater than 50 % of the jojoba oil once it is extracted (i.e., cold pressed). The waxy esters in the jojoba oil are chemically processed in the same manner as regular vegetable oil sources that contain triacylglycerides (or fatty acids). In other words the jojoba oil becomes heated and then gets mixed together with methanol and a catalyst such as potassium hydroxide. The result of this process produces both fatty alcohols (at ~ 20 %) and methyl esters (at ~ 80 %) as shown in Figure 4-11. The methyl esters, as mentioned above, are the biodiesel component of this process. The structure of the waxy ester in the original jojoba oil and the resulting jojoba fatty alcohol or jojoba methyl ester are shown in Figure 4-11.

The compounds derived from the jojoba oil can be utilized in a large variety of other products, especially health care or cosmetic types. The slight modification of the jojoba oil imparts special properties to these types of products. Some modified compounds include hydrogenated jojoba oil, jojoba alcohols, jojoba butter or ethoxylated jojoba oil.[195] All are an integral part of the cosmetic industry due to their special properties. Such properties include its use as an emoliant or as an antioxidant agent. Jojoba based compounds are especially known for their ability to easily penetrate the skin. Therefore, jojoba based cosmetic products include body creams, skin moisturizers, hair care products, various body oils, lotions and lipstick.[196]

Other product uses of jojoba include animal and human foods, gardening products, pharmaceuticals and lubrication or polishing materials.[197] The lubricants can be applied to heavy machinery or used as transformer oil (also mentioned previously for palm oil). The extracted seed cake is also valuable for the pharmaceutical industry or for application as natural pesticides for plants. Therefore, the harvesting of jojoba can yield both biodiesel and products of higher monetary value used as ingredients in cosmetic, industrial and healthcare products.

Salicornia is another desert oilseed that can be grown for processing into biodiesel or other biofuel manufacture. Several companies have been interested in exploring its potential use towards making bio-based jet fuel.[198,199] Salicornia contains both a high oil (~ 30 %) and protein content (~ 30 %). Therefore it tends to be suitable for biodiesel production as well as a healthy food source due to the different types of salts that accumulate within the plant. The plant is also known to contain beneficial pharmacological anti-inflammatory, anti-microbial and anti-oxidative properties.[201]

This plant grows very well in high saline sandy, silt, clay or marshy conditions. It can potentially be cultivated in desert areas or marshes that are located close to an ocean. Most importantly, it can be irrigated with seawater. In fact experiments have been done in the past where this oilseed was cultivated in Mexico alongside the sea of Cortez and it grew very well in just sandy soil with the addition of seawater.[200] Salicornia as a crop tolerates very high salt concentrations (NaCl) but also requires the addition of a nitrogen source in the soil. In addition, this oilseed crop is similar to cottonseed where the plant gets processed in a mill. At the mill the oil content from the seed normally gets extracted with a solvent like hexane.

Other desirable oilseed plants that grow in the desert include jatropha and castor beans. Both of these plants are effective oil producers. These oilseeds contain around 50 % oil in the seed itself and give large oil yields per acre. The castor bean mainly contains around 90 % ricinoleic acid while jatropha has a combination of palmitic, oleic and linoleic fatty acids. The castor bean plant also contains a very toxic ingredient in the seeds called ricin. So some care is needed when handling and processing it. Castor bean plants are moderate in height as they grow from 3 – 10 feet while jatropha trees are larger in size, growing up to 20 feet.[202,203] Jatropha trees on the average have a lifespan of around 50 years.

The similarities of cultivation between castor beans and jatropha

Grows well w/ moderate temperatures, Can only tolerate small amounts of frost conditions
Has little to moderate fertilizer requirements (Nitrogen, Phosphorus, Postassium – NPK)
Grows on marginal lands & has ability To rehabilitate the land
Needs well drained soil
Around ~ 50 % oil in the seed
Requires limited rainfall and can Withstand some drought
Grows in tropical, semi-arid And desert climates

Table 4-7 : Castor beans and Jatropha tree have many soil related growth conditions in common that allow it to grow well in the desert environment. In addition, the allowable temperature and rainfall amounts are similar so that these species of oilseeds can also grow in a number of similar regional areas of the world.

These plants exhibit similar growth conditions regarding temperature, soil and rainfall requirements as outlined in Table 4-7.[204,205] They are also known to live in similar regional locations such as tropical and semi-arid climates. Therefore, both plants can be grown in areas with limited rainfall. They can even withstand drought conditions for a certain period of time. These plants do not grow in cold climates and therefore, can tolerate only minimal amounts of frost conditions. Both plants require well drained soils yet need little to moderate amounts of fertilizer that has nitrogen, phosphorus and potassium. In addition, these plants can be utilized on marginal lands where they have the potential to rehabilitate the area.

Castor bean oil, similar to some of the other oilseeds (palm oil, jojoba oil) tends to be a common ingredient found in a number of consumer and industrial products such as lubricants, paints, coatings, nylon, adhesives, plastics, dyes, inks, cosmetics and soaps.[206,207] In summary, a number of oilseeds plants can be cultivated in the desert to produce oil for biodiesel as well as many other consumer and industrial products. These types of products have good economic value of high demand. Therefore, it would be both economically and environmentally prudent to increase cultivation of these types of desert oilseeds for multiple industrial purposes.

4.11 Improvements in algae cultivation may allow biodiesel production to take place more economically

The current cost of manufacturing biofuel or biodiesel from algae tends to be relatively high due to multiple processing steps that are energy intensive. These processing steps are necessary in order to separate the algae from solution, extract the lipids and convert them into biodiesel. The types of processing methods utilized for algae based biodiesel production include filtration, centrifugation, drying, size reduction, solvent extraction and esterification.[208] The specific processing steps of centrifugation, drying and size reduction are very energy intensive when applied.

The combination of utilizing these processing steps coupled with their high energy requirements tends to raise the cost of biodiesel production from algae. A scientific study published in 2011 asserted that the biodiesel cost per gallon made from algae can vary anywhere between $10 - $20 per gallon.[209] However, as improved, less energy intensive processes involved with harvesting and separation are developed, the cost of biodiesel production per gallon should also drop dramatically as well. Lower costs of production can also arise from taking advantage of renewable wastes emitted by an industrial process. Such an example includes harnessing the waste heat from a related manufacturing process in the refinery. This waste heat can be used to dry the algae.

Another factor that influences the cost of production deals with the manner in which the algae are cultivated. Processing costs can be reduced through certain cultivation methods. These methods should allow for the easier extraction of lipids taken from the algae. Easier extraction of the lipid material usually equates to less involved processing steps that are necessary. This helps to save on labor and energy requirements.

At least two types of potential algae cultivation methods could accomplish this feat as shown in Figure 4-12. One improved cultivation method involves gathering the lipids in higher concentrations within the cells. Growing intracellular lipids at high concentrations within the algae should assist towards making the extraction process more straightforward and easier.

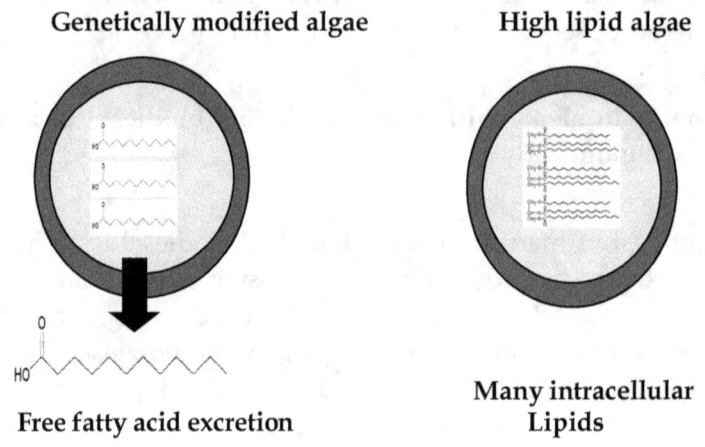

Figure 4-12 : Algae cultivated in different ways helps to determine how it is further processed into biofuel or biodiesel. Algae that emit free fatty acids or algae that produce high amounts of intracellular lipids work towards manufacturing biodiesel using easier processing methods.

A second cultivation method that greatly simplifies subsequent lipid extraction involves growing algae that excrete fatty acids into the solution media. This takes place with algae that are genetically modified or that are grown in certain growth conditions such as co-cultivation with bacteria.[210,211] After the fatty acids are excreted, methods such as solvent extraction can recover them. This may be able to take place by applying the correct solvent solution that circulates close to or near the growth media. This type of processing is much less involved compared to the several types of energy intensive steps described earlier in this section. However, this method has obstacles to overcome such as preventing toxicity effects from the solvents themselves. Another disadvantage includes the low overall amount of compounds excreted from the algae within a given amount of time. Such methods will then require improvements in the yield and rate at which fatty acids or other compounds are made.

Regardless, it is felt that improved cultivation methods will be the key towards lowering processing costs that allow algae to become an integral part of biodiesel production. Our further need for biodiesel may require more large scale production of certain feedstock sources in the future. Development of improved cultivation and processing methods for the conversion of algae into biodiesel would be one of the best ways to help us meet this goal.

REFERENCES:

143. Biosynthesis of biodiesel without glycerine byproduct – (Technology #z07163) by University of Minnesota – Office of Technology Commercialization – accessed online

144. Efficacy of specific gravity as a tool for prediction of biodiesel-petroleum diesel blend ratio – Fuel Vol 99 pgs 254 – 261 [2012] by B. Moser

145. Biodiesel : A realistic fuel alternative for diesel engines – Chapter 6 Fuel properties of biodiesels & Chapter 7 Current technologies in biodiesel production by A. Demirbas

146. Analysis and comparison of performance and emissions of an internal combustion engine fuelled with petroleum diesel and different biodiesels – Fuel vol 90 pgs 2147 – 2157 [2011] by P.McCarthy, MG Rasul, S. Moazzem

147. The effect of biodiesel and bioethanol blended diesel fuel on nanoparticles and exhaust emissions for CRDI diesel engine – Renewable Energy Vol 35 pgs 157 – 163 [2009] by H. Kim, B. Choi

148. Biodiesel handling and use guidelines - DOE/GO-102006-2358 by the US DOE EERE [2006]

149. Biodiesel production estimates 2005 – 2011 – Biodiesel Fact Sheets National Biodiesel Board – http://www.biodiesel.org/ -- - accessed May 2014

150. Diesel fuel technical review - [2007] Chevron Corporation – J. Bacha, J. Freel, A. Gibbs et al – http://www.chevronwithtechron.com/products/documents/Diesel_Fuel_Tech_Review.pdf

151. EPA Extramural Research – Final Report : Technology for enhanced biodiesel economics – http://cfpub.epa.gov/..

152. **Value added uses for crude glycerol – a byproduct of biodiesel production** – Biotechnology for Biofuels Vol 5 : 13 [2012] by F. Yang, MA Hanna, R. Sun

153. **The future of glycerol : new uses of a versatile raw material** – by M. Pagliaro, M. Rossi [2008] – Cambridge Royal Society of Chemistry

154. **Top value added chemicals from biomass – Vol 1 Results of screening from potential candidates from sugars & synthesis gas** – PNNL & NREL [2004] by T. Werpy & G. Petersen

155. **From glycerol to value added products** – Angewandte Chemie International Edition Vol 46 n 24 pgs 4434 – 4440 [2007] by M. Pagliaro, R. Ciriminna, H. Kimura et al

156. **The glycerine glut & options for the value-added conversion of crude glycerol resulting from biodiesel production** – Environmental Progress Vol 26 n 4 pgs 338 – 348 [2007] by DT Johnson & KA Taconi

157. **Glycerol feedstock utilization for oil based fuel manufacturing** – US Patent Application 20090148918 by DE Trimbur, et al [2009] – cited from USPTO patent application full text and image database

158. **Biodiesel production technology** – NREL/SR-510-36244 [2004] by J. Van Gerpen, B. Shanks, R. Pruszka et al – http://www.nrel.gov/docs/fy04osti/36244.pdf - accessed May 2014

159. **Biodiesel process equipment and systems** – Pacific Biodiesel Inc. – http://www.biodiesel.com/technologies/process-equipment-systems/ - accessed May 2014

160. **Production process [Biox corporation]** – http://www.bioxcorp.com/production-process/ - accessed May 2014

161. **A review on biodiesel production using catalyzed transesterification** – Applied Energy Vol 87 pgs 1083 – 1095 [2010] by DYC Leung, X Wu, MKH Leung

162. **Fast one phase oil rich processes for the preparation of vegetable oil methyl esters** – Biomass and Bioenergy Vol 11 pgs 43 – 50 [1996] by DGB Boocock, SK Konar, V Mao, H Sidi

163. **Revolutionary ultrasonics technology** – Genuine Bio-Fuel Inc. – http://www.genuinebiofuel.com/revtechnology.html - accessed May 2014

164. **Ultrasonic biodiesel reactor** – IncBio – http://www.incbio.com/ultrasonic_biodiesel_reactor.html - accessed May 2014

165. **Biodiesel production using enzymatic transesterification – current state and perspectives** – Renewable Energy Vol 39 No 1 pgs 10 – 16 [2012] by A. Gog, M. Roman, M. Tosa et al

166. **Biodiesel production with immobilized lipase : A review** – Biotechnology Advances Vol 28 pgs 628 – 634 [2010] by T. Tan, J. Lu, K. Nie, L. Deng et al

167. **Evaluation of diatomaceous earth supported lipase sol gels as a medium for enzymatic transesterification of biodiesel** – Journal of Molecular Catalysis B : Enzymatic Vol 77 pgs 92 – 97 [2012] by SM Meunier, RL Legge

168. **Use of a low cost methodology for detoxification of castor bean waste and lipase production** – Enzyme and Microbial Technology Vol 44 pgs 317 – 322 [2009] by MG Godoy, MCE Gutarra, FM Macial, SP Felix et al

169. **New Holland approves B100** – www-WallacesFarmer-com – Technology/Machinery p47 [Dec 2007] – http://magissues.farmprogress.com/WAL/WF12Dec07/wal053.pdf - accessed May 2014

170. **Case IH expands B100 biodiesel use in farm equipment** – Media → News Releases – http://www.caseih.com/.. - accessed May 2014

171. **Using biodiesel in John Deere engines** – [engines and drive trains section → Biodiesel] – http://www.deere.com/- accessed May 2014

172. **Biofuel Advances** – [New Holland] – http://agriculture.newholland.com/us/en/About-New-Holland/Innovation/Pages/Biofuel-Advances.aspx -- accessed May 2014

173. **September 2012 Flash report United States unit retail sales** – Ag Connect 2013 Expo – Association of Equipment Manufacturers – http://www.agweb.com/assets/1/6/12_9_USAG.pdf - accessed Mar 2013

174. **The California diesel fuel regulations** – Title 13 & 17 California code of regulations [2004] by California Air Resources Board

175. **Key properties and blending strategies of hydrotreated vegetable oil as biofuel for diesel engines** – Fuel Processing Technology Vol 92 pgs 2406 – 2411 [2011] by M. Lapuerta, M. Villajos, JR Agudelo, AL Boehman

176. **Co-processing methane in high temperature steam gasification of biomass** – Bioresource Technology Vol 128 pgs 553 – 559 [2013] by AW Palumbo, EL Jorgensen, JC Sorli, AW Weimer

177. **Biofuels from continuous fast pyrolysis of soybean oil : A pilot plant study** – Bioresource Technology Vol 100 No 24 pgs 6570 – 6577 [2009] by VR Wiggers, HF Meier, A Wisniewski et al

178. **Biodiesel : The sustainability dimensions** – National Sustainable Agriculture Information Service – ATTR [2010] by A. Kurki, A. Hill, M. Morris – http://attra.ncat.org/-

179. **Biofuels – Renewable energy sources : A review** – Journal of Dispersion Science and Technology Vol 31 pgs 409 – 425 [2010] by I. Kravola, J. Sjoblom

180. **Castor oil biodiesel and its blends as alternative fuel** – Biomass and Bioenergy Vol 35 pgs 2861 – 2866 [2011] by P. Berman, S. Nizri, Z. Weisman

181. **Improving the low temperature properties of biodiesel : Methods and consequences** – Renewable Energy Vol 35 pgs 1145 – 1151 [2010] by PC Smith, Y Ngothai, QD Nguyen

182. **Preparation and evaluation of jojoba oil methyl esters as biodiesel and as a blend component in ultra-low sulfur diesel fuel** – Bioenergy Research Vol 3 No 2 SI pgs 214 – 223 [2010] by SN Shah, BK Sharma, BR Moser, SZ Erhan

183. **Handbook of Energy Crops** – by JA Duke [1983] – http://www.hort.purdue.edu/newcrop/duke_energy/dukeindex.html - accessed May 2014

184. **Energetic and economic feasibility associated with the production and conversion of beef tallow to a substitute diesel fuel** – Biomass and Bioenergy Vol 30 pgs 584 – 591 [2006] by RG Nelson, MD Schrock

185. **Review of biodiesel composition, properties and specifications** – Renewable and Sustainable Energy Reviews Vol 16 No 1 pgs 143 – 169 [2012] by SK Hoekman, A Broch, C Robbins et al

186. **Biodiesel and renewable diesel : A comparison** – Progress in Energy and Combustion Science Vol 36 pgs 364 – 373 [2010] by G. Knothe

187. **Evaluation of nine oil crops for fatty acid constituents of their oils** – Journal of King Abdulaziz University : Meteorolgy, Environment and Arid Land Agriculture Sciences Vol 22 No 1 pgs 51 – 59 [2011] by FS El-Nakhlawy, MS Shiboob

188. **Characteristic and composition of Jatropha curcas oil seed from Malaysia and its potential as biodiesel feedstock** – European Journal of Scientific Research Vol 29 No 3 pgs 396 – 403 [2009] by E. Akbar

189. **Determination of tri-saturated glycerides in lard, hydrogenated lard and tallow** – Oil and Soap Vol 23 pgs 385 – 389 [1946] by FE Luddy, RW Riemenschneider

190. **Industrial Processes & the Environment – Crude Palm Oil Industry** – Ch 2 – The crude palm oil industry – an overview : Ch 4 – Environmental Issues – Editors MI Thani, R. Hussin et al [1999]

191. **Production of activated carbon from oil palm empty fruit bunches for removal of zinc** – 12th International Water Technology Conference [2008] by MZ Alam, SA Muyibi, N Kalmaldin

192. **Lignocellulosic fiber media filters as a potential technology for primary industrial wastewater treatment** – Jurnal Teknologi Vol 49 pgs 149 – 157 [2008] by MGM Nawawi, N Othman, AN Sadikin et al

193. **Dielectric properties of palm oils as liquid insulating materials : effects of fat content** – Proceedings of 2005 International Symposium on Electrical Insulating Materials Vol 1 pgs 91 – 94 by S. Aditama

194. **Biotechnological Production of Plant Secondary Metabolites** – Editor I. Orhan [2012] Betham Books

195. The International Jojoba Export Council – http://www.ijec.net/jojoba-facts - Jojoba Facts - The Jojoba Market – accessed May 2014

196. The International Jojoba Export Council - http://www.ijec.net/manufacturing-with-jojoba - Manufacturing with Jojoba - Applications - accessed May 2014
197. **The struggles of jojoba** - Chemtech 1995 pgs 49 - 54
198. **Salicornia producer preps for delayed Interjet biofuel demo** - by M. Kuhn [Jan 2010] - http://www.flightglobal.com/ -- accessed May 2014
199. **Sustainable aviation fuel** - http://www.airbus.com/ -- Home / Innovation / Future by Airbus - accessed May 2014
200. **Water requirements for cultivating Salicornia Bigelovii tori with seawater on sand in a coastal desert environment** - Journal of Arid Environments Vol 36 pgs 711 - 730 [1997] by E. Glenn, S. Miyamoto, D. Moore et al
201. **Salicornia herbacea : Botanical, chemical and pharmacological review of halophyte marsh plant** - Journal of Medicinal Plants Research Vol 3 no 8 pgs 548 - 555 [2009] by MH Rhee, HJ Park, JY Cho
202. **Jatropha propagation** - by R. Hoyt - http://homeguides.sfgate.com/jatropha-propagation-22162.html -- accessed May 2014
203. **Castor bean** - http://www.oilseedcrops.org/castor-bean/ - accessed May 2014
204. **Castor bean - A fuel source for the future** - Centre for Jatropha Promotion - http://www.jatrophaworld.org/castor_bean_83.html -- accessed May 2014
205. **The plant profile** - Centre for Jatropha Promotion - http://www.jatrophaworld.org/jatropha_plant_9.html -- accessed May 2014
206. **High tech castor plants may open door to domestic production** - USDA - ARS - http://www.ars.usda.gov/is/ar/archive/jan01/plant0101.html - accessed May 2014
207. **Exploiting EST databases for the development and characterization of EST-SSR markers in castor beans** - BMC Plant Biology 10:278 [2010] by L. Qiu, C. Yang, B. Tian et al
208. **Chapter 10 Downstream processing of cell-mass and products** - by E. Molina Grima, FG Acian Fernandez, A Robles Medina - **Handbook of microalgal culture : Biotechnology and applied phycology** - by Blackwell Publishing LTD [2004]
209. **Techno-economic analysis of autotrophic microalgae for fuel production** - Applied Energy Vol 88 no 10 pgs 3524 - 3531 [2011] by R. Davis, A. Aden, PT Piankos
210. **Chlorellin, an antibacterial substance from chlorella** - Science Vol 99 No 2574 pgs 351 - 352 [1944] by R. Pratt, TL Daniels, JJ Eiler et al

211. Genetic engineering of algae for enhanced biofuel production – Eukaryotic Cell Vol 9 No 4 pgs 486 – 501 [2010] by R. Radakovits, RE Jinkerson, A. Darzins

Chapter 5

Alternative Ethanol Production Methods

5.1 Alternative ethanol technologies are superior manufacturing methods when compared to conventional fermentation

This publication encourages alternative ethanol production over corn starch ethanol manufacture. In the future the manufacture of ethanol may very well emanate from alternative technologies other than conventional fermentation. The alternative technology of conventional lignocellulosic fermentation has similar production related aspects in common with corn starch ethanol production. Therefore, as a method of producing ethanol, lignocellulosic fermentation suffers from certain drawbacks in application as well. Two of these drawbacks include inefficient pretreatment processes and the lack of a consolidated bioprocessing method. Consolidated bioprocessing basically combines the methods of saccharification and fermentation in one process using microbes that can accomplish both tasks.

Developing alternative ethanol production technologies promise to be superior to either of these conventional fermentation methods (lingocellulosic and corn starch) in almost every facet of production. These newer technologies are more energy efficient as well as better at consuming bio-resources in a more practical manner. They offer the advantages of high ethanol yield per ton of biomaterial processed, lower greenhouse gas (GHG) emissions, improved energy production through recycling of materials and the manufacture of alternative chemical byproducts. In addition, alternative ethanol production tends to be conformable to a wide range of feedstocks. These alternative ethanol production methods have the capability to handle several types of feedstocks at one time. Alternative ethanol technologies are also cost competitive when compared with other types of fuel manufacturing.

These particular aspects qualify them as technologies that are associated with the Integrated Biorefinery production model. The alternative ethanol production methods reported in this chapter revolve around the effective design of integrated biorefineries that incorporate more modern technologies. Four alternative ethanol technologies will be covered in this chapter. They include 1) **algae based assimilation of carbon dioxide**, 2) **fermentative acetic acid conversion**, 3) **synthesis gas fermentation** and 4) **thermochemical gasification**.

Algae based ethanol production, the first of these technologies to be discussed, has received attention in recent years due to advanced technological developments. Previous manufacturing methods of ethanol from algae gave low yields and rates of production that made it impractical for wide scale use. This picture has changed with the advent of genetically modified algae that produce higher amounts of ethanol. In fact, this model has been so successful that other types of alcohols, in addition to ethanol, are beginning to be produced from genetically modified microbes and algae. Chapter 7 discusses more detailed aspects involved with alcohol production from algae.

A second alternative technology, fermentative acetic acid conversion, has the potential to be the most successful lignocellulosic based ethanol production method to come about recently. It gives the highest yield of ethanol per ton of lignocellulosic biomass by efficiently utilizing all of the biomaterial provided. The final two technologies to be discussed, synthesis gas fermentation and thermochemical gasification, are alternative ethanol technologies that have some aspects in common. The principles of thermochemical gasification have already been covered in chapter 3. However, ethanol based thermochemical gasification has already been developed towards processing most types of municipal wastes for ethanol manufacture. The fourth technology, synthesis gas fermentation, combines both thermochemical gasification and fermentation, all of which take place within the same facility. It offers many advantages over other types of production methods including the ability to incorporate a wide range of lignocellulosic feedstock types.

5.2 Alternative ethanol production methods are more cost effective and produce more ethanol per ton of biomass

Table 5-1 compares the four alternative ethanol technologies to conventional lignocellulosic fermentation. The table presents the case that these newer methods are superior in production cost per gallon and yield more ethanol per ton of biomass or carbon dioxide provided. The data from the table also associates each production method with a company executing a particular technology. Such examples include two companies (***INEOS Bio*** and ***Coskata***) that in the near future are planning to produce ethanol using synthesis gas fermentation.[212,213,214] Other companies have developed their own proprietized technologies as well. ***Zeachem*** manufactures ethanol from a hybrid acetic acid fermentation method.[215] ***Algenol*** produces ethanol from carbon dioxide assimilation through their Direct to Ethanol® production method.[216,217] ***Enerkem*** is another integrated biorefinery company that makes ethanol through the thermal gasification of municipal waste.[218]

These alternative ethanol manufacturing companies produce larger amounts of ethanol per ton of biomass utilized when compared to conventional lignocellulosic fermentation. Conventional lignocellulosic ethanol fermentation only manufactures around 72 gallons of ethanol per ton of biomass, the lowest yield of all methods.[219] This result was obtained by averaging the yield of four lignocellulosic fermentation refineries from the DOE integrated biorefinery program. On the other hand, the alternative ethanol manufacturing methods of fermentative acetic acid, synthesis gas fermentation, thermochemical gasification and algal carbon dioxide assimilation all manufacture much higher yields of ethanol per ton of material processed. Themochemical gasification of biomass manufactures greater than 90 gallons of ethanol per ton.[218] Similarly, synthesis gas fermentation produces just over 100 gallons per ton.[214] Even higher yields are expected from fermentative acetic acid conversion (135 gallons per ton)[215] and algal based carbon dioxide assimilation (~ 125 gallons per ton of carbon dioxide).[216]

Estimated ethanol cost of production and yield from various manufacturing methods

Production Method	Company or Institution	Estimated Cost per Gallon	Ethanol yield Per ton Biomass
Lignocellulosic ethanol fermentation	DOE Integrated Biorefinery Program **	-----	~ 72 gallons / ton**
Synthesis gas fermentation	Coskata	~ $ 1.00	> 100 gallons / ton
Synthesis gas Fermentation	INEOS Bio	Cost competitive	~ 105 gallons / ton
Thermochemical Gasification	Enerkem	> $1.00	> 90 gallons / ton
Fermentative acetic acid conversion	Zeachem	< $1.00	135 gallons / ton
Carbon dioxide assimilation	Algenol	> $1.00	~ 125 gallons / ton

Table 5-1 : The expected cost of ethanol production and ethanol yield per ton of biomass provided from various ethanol manufacturing methods that include fermentation, mixed alcohol production and alternative production methods.[212,213,214,215,216,217,218,219,220,221]
** Ethanol average yield from 4 integrative lignocellulosic fermentation refineries

The cost required to make the ethanol also factors in highly when considering which production technologies are superior. Most of the alternative ethanol production technologies are very cost competitive. The technologies of synthesis gas fermentation (Coskata)[212], fermentative acetic acid conversion (Zeachem)[215], and algae based ethanol (Algenol)[216] all produce ethanol or other alcohols at around $1.00 or slightly higher per gallon. According to a US DOE case study performed in 2010, ethanol made from thermal gasification is projected to cost over $1.00 per gallon and varies a bit according to the size of the refinery.[221]

Aside from producing large yields of ethanol and being cost effective, these alternative technologies are also efficient in other ways related to production. For example, they emit less amounts of carbon dioxide and require modest energy requirements for production. A more in depth description of the basic manufacturing operating principles that allow these technologies to excel in efficiency are discussed further on. The production technologies of algae based ethanol, fermentative acetic acid conversion, synthesis gas fermentation and thermal gasification are discussed in this respective order over the remaining sections of this chapter.

5.3 Genetic engineering allows seawater algae (cyanobacteria) to produce ethanol in ample quantities

It has been known for decades that seawater algae (cyanobacteria) produces ethanol when cultivated as a 'microbial mat' as shown in Figure 5-1.[222] The production of ethanol executed in this manner takes place through two separate processes that, respectively, occur during the day and then during the darker nighttime hours. Photosynthesis during the day makes large amounts of polysaccharides (sugars) while respiration executed at nighttime breaks down these saccharides into ethanol. Algae that accumulate on mats floating on a water surface receive high amounts of sunlight and oxygen that are more easily distributed among the algae population. These types of conditions encourage the growth and accumulation of saccharides within the algae during the daytime. During the nighttime as respiration takes place, the saccharides become converted through metabolic pathways into various chemicals. These chemicals include organic acids, hydrogen and ethanol as shown in Figure 5-1.

However, researchers have found that cultivating natural seawater cyanobacteria in this manner creates very little amounts of ethanol worthy of being collected for biofuel end use. This, in part, is due to the fact that the saccharides are broken down into several end products other than just the ethanol. In addition, the conversion of saccharides into ethanol with the algae's available internal enzymes tends to be inefficient in process as well. On the other hand, it is known that microbes such as yeast cells convert most of their given saccharides into ethanol through the operation of their more efficient enzymes. Therefore, an alternative approach that allows algae to make large amounts of ethanol involves genetically modifying the algae so that they produce different enzymes other than the ones they already have. An insertion of a set of genes placed into the algae encode for these enzymes, which are originally taken from microbes such as yeast cells. This particular genetic engineering technique specifically involves taking the enzyme production genes from the microbe *Zymomonas mobilis* and then inserting them into the cyanobacteria (algae).[223,224]

The formation of ethanol and other products from the cultivation of cyanobacteria done on microbial mats

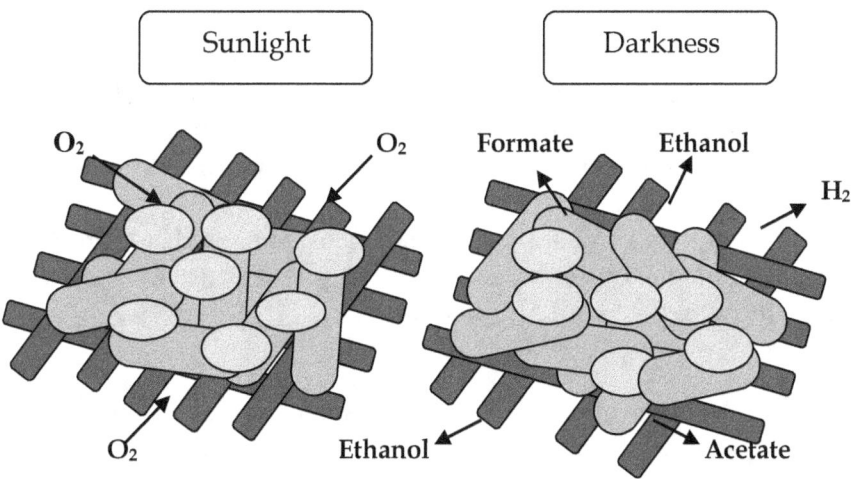

Figure 5-1 : The growth of cyanobacteria on microbial mats can produce ethanol, organic acids and hydrogen at night due to the amount of sugars that build up during the day that are attributed to a high amount of sunlight and oxygen.

The two enzymes that are essential for allowing the direct conversion of the metabolic intermediate pyruvate into ethanol are *pyruvate decarboxylase* and *alcohol dehydrogenase*.[223,224] Normally these enzymes are tied to saccharide based fermentation that happens to be commonly executed among microbes in general. However, thanks to the wonders of modern technology, these enzymes have been adapted to make ethanol from incoming carbon sources such as carbon dioxide. This bodes well for microorganisms such as algae. This type of genetically based revolutionary breakthrough allows other renewable waste sources such as carbon dioxide to be utilized towards future biofuel production.

5.4 The Algenol Process manufactures ethanol in photobioreactors using seawater & carbon dioxide

Genetically altered seawater algae containing these enzymes have the capability of producing greater amounts of ethanol when assimilating carbon dioxide. Through photosynthesis, carbon dioxide and water are processed together to help form the ethanol. The carbon dioxide does not directly turn into ethanol but rather it becomes converted into other metabolic intermediates that boost the formation of pyruvate. An accumulation of the pyruvate then subsequently gets processed by the genetically altered enzymes. This method makes ethanol more efficiently and with greater yield due to the use of these enzymes.

One company has already utilized this technology with genetically altered seawater algae in order to make ethanol in appreciable quantities. Algenol is the only algae based company manufacturing ethanol at a pilot plant level. They have been able to accomplish this in part through the DOE based integrated biorefinery demonstration grant. The Algenol pilot plant setup is very interesting in configuration. Table 5-2 provides more specifics concerning how much approximate land and photobioreactors are used to accomplish this type of ethanol production.

Approximate ethanol production, yield, land acreage and photobioreactors at the Algenol pilot plant facility

Approximate amount of Ethanol to be produced	100,000 Gallons per year
Yield per acre	6,000 Gallons per acre [improved to 8,000 gallons per acre]
Approximate land use	~ 17 acres
Carbon dioxide use	~ 720 tons (2 tons per day)
Approximate number of Photobioreactors	~ 2900 photobioreactors

Table 5-2 : Total ethanol production, yield per acre, land use and number of bioreactors at the Algenol pilot plant facility in Freeport, TX.[216,225,226]

This integrated biorefinery facility is estimated to contain around 2900 photobioreactors on 17 acres of land.[225] Estimates in production from the pilot plant project that 100,000 gallons of ethanol are made per year using approximately 720 tons of carbon dioxide.[226] This gives rise to an excellent yield of ethanol produced per acre. The pilot plant projects manufacturing ethanol at a yield of 6,000 gallons per acre. However, recently the process has so improved to the point to where the company is now projecting that yields around 8,000 gallons of ethanol per acre are expected.[217] This yield per acre is outstanding when compared to other forms of ethanol production such as crops or other plant sources.

The process of making and harvesting the ethanol performed by Algenol is novel in application. The ethanol collection and harvesting process takes place within their photobioreactors as shown in Figure 5-2. The plastic photobioreactors are shaped like long semicircle cylindrical objects. Their method of ethanol production is simple in concept. The cyanobacteria are cultivated within the photobioreactors where they excrete the ethanol in appreciable quantities and do so at a very good rate. Through evaporation, water vapor and ethanol condense at the top surface of the photobioreactors and then drain to the bottom through troughs that collect both the ethanol and water.[227] The ethanol-water mixture then travels down the troughs upon a mechanical displacement of the photobioreactors. Once the mixture has been collected, another downstream processing technology is applied. This includes vapor compression steam stripping to further separate and concentrate the ethanol away from the water.[227]

A set of plastic photobioreactors that cultivate algae and collect ethanol and water using troughs

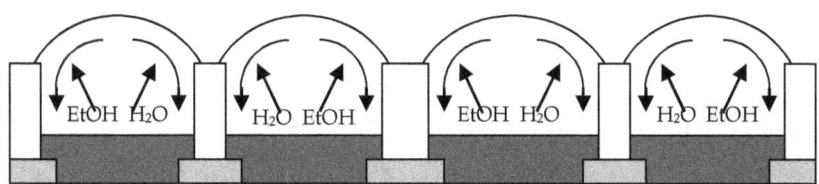

Figure 5-2 : Representation of a set of photobioreactors used to cultivate algae in seawater that is supplemented with carbon dioxide. Water and ethanol vaporize within the photobioreactor and collect within troughs at the sides of the reactors.

The above mentioned process also helps to produce purified water for potential consumer and industrial use. An excellent yield of two gallons of fresh water are made for every gallon of ethanol manufactured.[216] The idea of producing fresh water as a byproduct from seawater is a desirable concept. It's impressive that this type of facility has the capability of making a large amount of fresh, clean & potable water as a co-product. The commercial and environmental applications of such a process are far reaching, especially in the sandy desert areas that line many seashores. The production of fresh purified water will allow other businesses as well as residences to arise in locations close to the algae ethanol refinery.

In order for the genetically modified algae to successfully emit ethanol during growth, they must become acclimated to the intense light and heat conditions of the outside environment. Therefore, the algae are initially cultivated under varying light and temperature conditions in greenhouses so that they become accustomed to the desert environment where they will later flourish and grow in the photobioreactors. This is shown in Figure 5-3. During the acclimation process, algae are initially grown in greenhouses with less light and then are moved into other greenhouses that have higher light intensity. This type of process allows the algae to gradually get used to high temperature and light intensive conditions of the desert. This selective process of cultivation allows for more robust, durable algae species to become acclimated to their hot desert environment. In order for this to happen, it is necessary to cultivate sets of genetically altered and wild type algae together in the greenhouses. The process of hybridization allows wild normal species of algae to become genetically altered through association and interaction with the genetically modified algae. Hybridization takes place when algae share/transfer genetic material between each other, resulting in superior growth qualities as shown in part B of Figure 5-3.

Cyanobacteria become hybridized in greenhouses in order to become acclimated to desert conditions

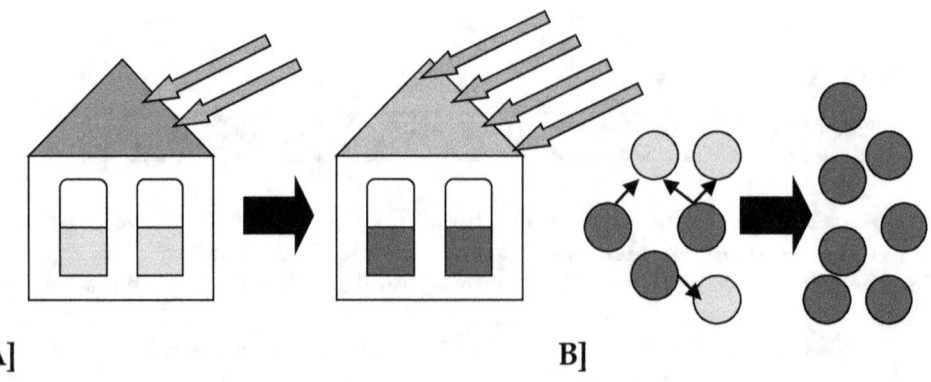

A]   B]

Figure 5-3 : A] The production of algae grown under varying condition of light and heat within greenhouses. B] The sharing of genetic information between algae species is done through the process of hybridization.

5.5 Ethanol produced from algae has a higher net energy value, higher yield per acre and improved greenhouse gas emissions

Two important factors related to the efficacy of biofuel production are the net energy contained in the biofuel after producing it as well as the amount of carbon dioxide emitted during the process. The metrics that better describe and quantify these two factors are Net Energy Value (NEV) and Greenhouse Gas (GHG) Emissions. As previously mentioned in chapter 1, NEV is calculated as the difference between the inherent energy content of the fuel and the energy input required to produce it. Therefore, the NEV values of different production methods can be directly compared with one another. On the other hand, the GHG values from various biofuel production methods are directly compared to petroleum vehicle fuel manufacturing in order to ascertain whether they are environmentally friendly. Therefore, GHG emission values for various biofuel production methods are compared with one another based upon their original comparison with petroleum vehicle fuel GHG emissions. The Algenol production process (Direct to Ethanol®) excels in both these metrics (NEV & GHG) when compared to corn ethanol based production.

The NEV values of corn based ethanol versus the NEV value projected from the Algenol process are shown in Figure 5-4. Corn based ethanol has a Net Energy Value (NEV) that varies between 10 – 15 %[228] while the Algenol process has an estimated NEV in the neighborhood of 60 %.[225] A higher NEV denotes that more leftover energy is available in the biofuel for future utilization. Therefore, Algenol's return in energy for ethanol appears to be 3-6 times higher when compared to corn ethanol NEV. The results also demonstrate that the Algenol production process is not as energy intensive, i.e., does not require as much energy to produce, as corn starch production. In addition, the excellent NEV value of the Algenol production process is comparable or similar to other hybrid (alternative) ethanol production processes. The Algenol NEV value was calculated by the author based on literature information concerning the Algenol process.[225] The amount of energy used to produce ethanol during manufacturing includes fertilizer use, vapor compression and stripping, heat generated, transportation, pumps, mixing and other related processes.

Carbon dioxide based GHG emissions are also more favorable with the Algenol production process when compared to both petroleum and corn starch ethanol manufacture. Estimates project that Algenol's process gives at least 20 % or higher reduction in GHG emissions when compared to petroleum manufacture.[225] Corn starch ethanol production also shows around a 20 % decrease in GHG emissions. However, it can be shown that there is a possibility to increase the reduction of GHG emissions from algae ethanol to an overall 80 % reduction value.[225] This may be accomplished by utilizing certain types of alternative energy sources. For example, combined heat and power obtained from natural sources such as biomass would allow for such a significant reduction in GHG emissions.

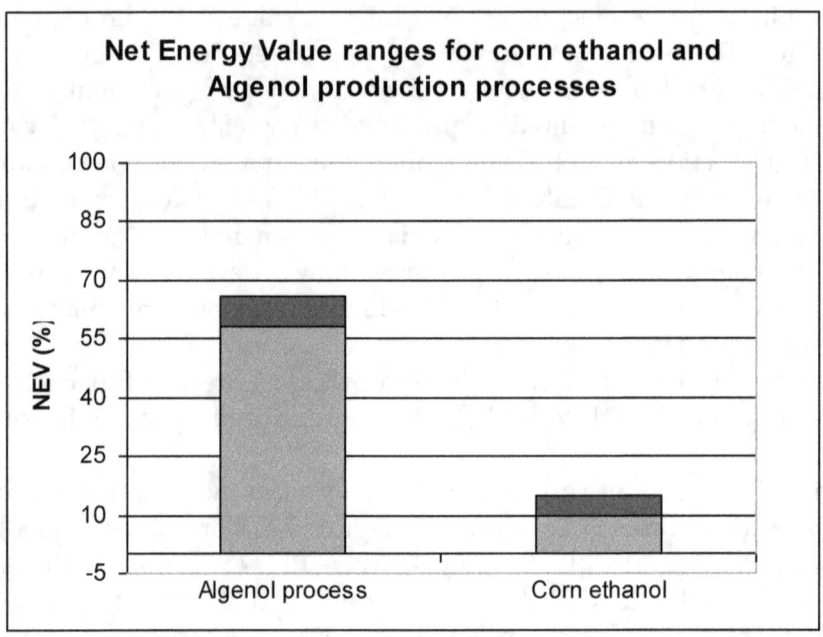

Figure 5-4 : Net Energy Value comparison and ranges between the production processes of corn ethanol and the Algenol based ethanol production. Algenol Net Energy Value calculated from power requirements of various estimated process stages of the Algenol process (i.e., estimated** not actual values).[225]

An important point of mention concerning overall GHG emissions has to do with the amount of carbon dioxide assimilated by the algae. Just as large or even a higher amount of carbon dioxide may be assimilated by the algae when compared to the amount emitted by manufacturing the ethanol. In other words, an algae based biofuel refinery has the ability to intake a large source of carbon dioxide flowing from another manufacturing plant. This amount of carbon dioxide taken in by the algae may be larger than the amount emitted by the manufacturing processes that produce ethanol from algae. Therefore, the assimilation of carbon dioxide from manufacturing processes such as these has the potential to contribute towards overall carbon negative GHG emissions. A carbon negative emission denotes that more carbon dioxide is absorbed by the algae than the net amount released by the other manufacturing processes of the refinery.

Another important factor associated with ethanol production from algae deals with its improved yield per acre when compared with energy or food crop based manufacture. Table 5-3 compares the yield of ethanol production per acre between algae ethanol, corn starch and switchgrass fermentation methods. The ethanol yield per acre is much higher for algae ethanol production than the other methods. Corn starch and switchgrass based ethanol fermentation yield 350 and 800 gallons per acre, respectively.[229,230] On the other hand, the Algenol ethanol production process has the potential to make around 8,000 gallons of ethanol per acre. This equates to at least 10 times or more the amount of ethanol made per acre of land utilized. Therefore, less land resources are needed for algae cultivation when compared to other energy or food crop cultivation needs. The land for the Algenol process is used to accommodate the photobioreactors and other equipment needed to produce the ethanol. Practical use of land resources becomes very important when considering the overall sustainability of biofuel production. Sensible cultivation of biofuel based feedstocks on appropriate land resources helps to achieve sustainability. Such considerations include algae cultivated on land that may not have to compete for other purposes like marginal or desert land areas.

Ethanol yield per acre comparing corn starch, switchgrass and the Algenol production process

Ethanol production method	Yield per acre
Corn starch ethanol	350 gallons per acre
Fermentation with switchgrass	800 gallons per acre
Algenol ethanol production	8,000 gallons per acre

Table 5-3 : Ethanol yields per acre compared between corn starch, switchgrass lignocellulosic production and the Algenol production process.[217,229,230]

5.6 There are many ways to produce acetic acid via fermentation

Acetic acid is one of the most common chemical products formed from fermentation. This fact infers that it should be possible to produce other types of chemicals from the acetic acid in a manufacturing environment after it has been isolated from the fermentation media. For example, the further thermochemical processing of acetic acid directly from fermentation can result in the production of ethanol. This takes place by adding another chemical with the acetic acid and then heating them together in order to allow the proper reaction to take place. Subsequent hydrogenation then converts the intermediate product into ethanol.

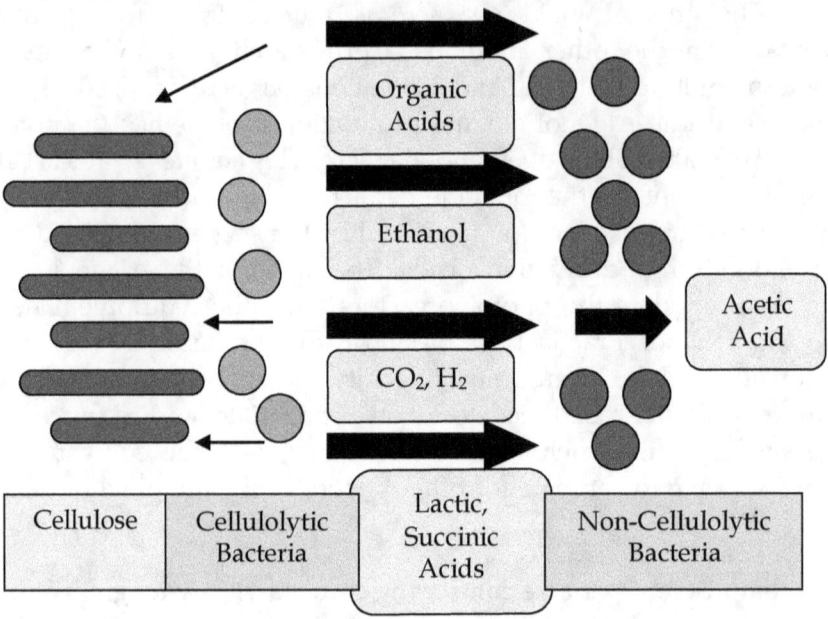

Figure 5-5 : Cellulolytic bacteria break down cellulosic material and convert the cellulose into many products while some of the non-cellulolytic bacteria further convert organic acids, ethanol, etc into acetic acid as the main product.[231]

What makes the manufacture of acetic acid from fermentation so special has to do with the large variety of microbes and methods that are available for this purpose. The microbes implemented for acetic acid fermentation can consist of a mixture of different types or contain specie(s) that are part of a specific class of microbes. The methods utilized to make the acetic acid from certain microbes are related to the form in which the saccharide material is provided to them. The material could consist of crude cellulose or broken down saccharides obtained from cellulose.

One common acetic acid fermentation method involves a mixture of microbes that are able to digest cellulose and then make the acetic acid. This can be executed as a two step process using a microbial mixture of cellulose degrading (cellulolytic) and non-cellulose degrading (non-cellulolytic) bacteria as shown in Figure 5-5.[231] The cellulolytic bacteria break down the cellulose material into the various products of organic acids, ethanol, hydrogen and carbon dioxide. The non-cellulolytic bacteria then convert these compounds mainly into acetic acid.

Another method utilizes species that are part of a group of microbes known as *Clostridia spp*. The cultivation of the species *clostridium lentocellum* is one such example. This species has the ability to digest cellulose in an available form. It tends to make a mixture of products such acetic acid, ethanol, carbon dioxide and hydrogen during fermentation involving cellulose.[232] This method also gives an excellent yield of acetic acid per gram of cellulose provided.

Outline of a process that makes acetic acid from glucose & carbon dioxide

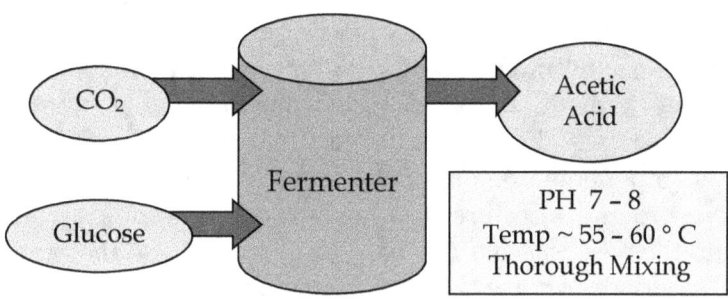

Figure 5-6 : Conditions of a process used to produce a high yield of acetic acid from glucose and carbon dioxide also using certain pH, temperature and mixing conditions.[233]

However, an even better yield of acetic acid can be obtained when the clostridium species metabolizes both glucose and carbon dioxide. One such example includes the cultivation of *clostridium thermocellum* as an industrial production model. The model incorporates a large reactor operated at conditions of pH 7 - 8, temperatures from 55 - 60 ° C and thorough mixing of carbon dioxide and glucose added to the system as shown in Figure 5-6.[233] The metabolic assimilation of both chemicals contributes towards making a higher amount of acetic acid. *Clostridium thermocellum* cultivated in this manner demonstrates how a class of microbes known as **acetogens** operates. Acetogens are the primary microbes used by the company Zeachem to make acetic acid as part of their proprietary ethanol manufacturing process. Their production model is described in more detail in the next section.

5.7 The company Zeachem manufactures ethanol efficiently through fermentative acetic acid conversion

The company Zeachem plans to manufacture ethanol as a hybrid production method that combines fermentation along with further thermochemical processing. The author describes this manufacturing method as <u>fermentative acetic acid conversion</u> since acetic acid is the primary fermentation product that gets converted into ethanol. This method compares similarly to the manufacturing method of carboxylate counter-current fermentation discussed in the next chapter. A depiction of the unit processes describing the operation of fermentative acetic acid conversion is outlined in Figure 5-7.

The intake process requires that the lignocellulosic material becomes pretreated before being passed to the fermentation stage. Zeachem plans to utilize the lignocellulosic feedstock source of hybrid poplar trees. Compared to other feedstocks, hybrid poplars grow very rapidly and can be harvested after 3-5 years of growth. A process called fractionation, shown in the figure, successfully separates the lignocellulosic material from the trees into its respective components of lignins, hemicellulose and cellulose. The fractionation process also breaks down the hemicellulose and cellulose portions into the basic saccharides of xylose and glucose that are required in this form as food for the acetogen microbes. Acetogens offer several advantages over other microbes that conduct fermentation on lignocellulosic material since they have the ability to metabolize both the cellulose and hemicellulose biocomponents (i.e., glucose & xylose) of it. This allows for more utilization of the biomaterial towards biofuel production as well as its improved carbon conversion during manufacture.

After fractionation and fermentation take place, the chemical acetic acid becomes converted into an ester called ethyl acetate through thermochemical combination of the acetic acid with ethanol. This chemical reaction is known as esterification. The ethyl acetate then gets further isolated and becomes hydrogenated in order to produce the end product of ethanol. Hydrogenation involves the addition of hydrogen in order to reduce and break apart the ester compounds into a set of two alcohols, which in this case are both ethanol. The hydrogen provided for the reduction process is obtained from the thermochemical gasification of lignin based residues formed during the initial intake and pretreatment of lignocellulosic material. Thermal gasification produces synthesis gas whereby hydrogen is the key component gas isolated from this process. The thermal gasification of lignin residues also assists in producing electrical power. This takes place by combusting some of the synthesis gas in order to produce steam required for the operation of steam turbines.

The Zeachem acetic acid conversion into ethanol production process

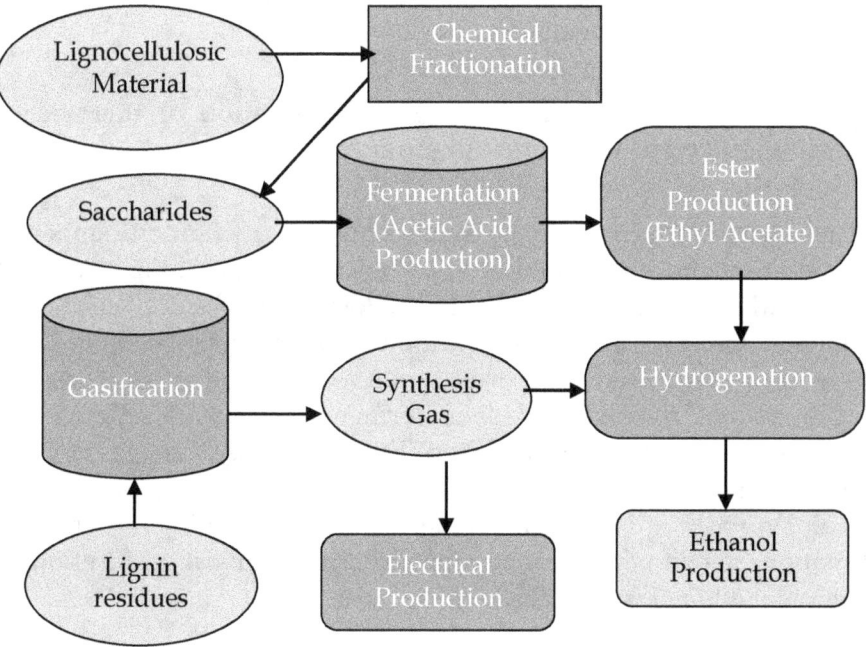

Figure 5-7 : The ethanol production process that manufactures ethanol from acetic acid using esterification and hydrogenation used by the company Zeachem.[234]

Other advantages are associated with utilizing the Zeachem fermentative acetic acid conversion model. This type of production model qualifies as integrated biorefinery technology because they make other chemical byproducts of interest in addition to ethanol. Such value added byproducts include acetic acid, ethyl acetate, ethylene, lactic acid, propionic acid, butanol, hexanol and hexane.[235] The Zeachem production method also achieves greater than 90 % reduction in GHG emissions when compared to petroleum manufacture.[236] These results were obtained from a life cycle analysis (LCA) study performed on the whole production process originating from feedstock cultivation to the manufacturing of the ethanol.[236] Aspects such as a high yield of ethanol per ton of fed biomass along with the non-emission of carbon dioxide from fermentation contributed to these results. In addition, Zeachem also claims that their production process is the most efficient type of ethanol manufacture due to its overall successful carbon conversion. Since the manufacturing process doesn't emit carbon dioxide, they assert that the carbon efficiency conversion is close to 100 %.[234] In summary, the hybrid ethanol manufacturing process executed by the company Zeachem has many advantages and benefits compared to other production technologies. High ethanol yield, efficient carbon conversion, low GHG emissions, assimilation of xylose and glucose sources, additional byproducts made and recycling of lignin residues towards electrical power and hydrogen generation will all contribute towards its potential success.

5.8 Synthesis gas fermentation is a great application of thermochemical gasification utilized towards ethanol production

The intriguing technology of synthesis gas fermentation happens to be another energy efficient ethanol production method that utilizes many types of lignocellulosic feedstocks. This technology involves the gasification of ligocellulosic materials and then continues with microbes feeding off of the synthesis gas components in order to make the ethanol. Synthesis gas produces ethanol through microbial fermentation using selected bacteria known as **anaerobic mesophiles**. These bacteria ferment the gas components of carbon monoxide, hydrogen and carbon dioxide into ethanol and other chemical byproducts.[237] The advantages of implementing a synthesis gas fermentation system versus conventional thermochemical gasification for the production of ethanol is listed below.

- Microbes can partially metabolize the carbon dioxide[237] component from the synthesis gas as well as tolerate low levels of gas impurities such as tars and hydrogen sulfide.[238]

- The cleanliness and consistency of the synthesis gas manufactured does not have to be as exact as that required for the thermal gasification process.

The bacteria that metabolize synthesis gas into ethanol prefer the carbon monoxide component of the synthesis gas. However, with higher gas pressures and good mixing conditions introduced into bioreactors, hydrogen and carbon dioxide also become assimilated by the bacteria.[238,239] The usual composition range of synthesis gas components provided to the bacteria are shown in Table 5-4. As shown in the table, low amounts of hydrogen sulfide and tars contained in the synthesis gas (around 1 %) can be tolerated by the microbes during the fermentation process. The percentage range of carbon monoxide, hydrogen and carbon dioxide provided to the microbes can vary a bit according to how the synthesis gas is produced. Usually carbon monoxide is in the greatest amount (40 – 65%), followed by hydrogen (25 – 35%) and then carbon dioxide (1 – 20%). Small amounts of methane (0 – 7%) also tend to show up in the synthesis gas.

The range of various gases & other impurities in synthesis gas that can be provided to microbial fermenters

Tars	Carbon Monoxide	Hydrogen	Methane	Carbon Dioxide	Hydrogen Sulfide
~ 1 %	40 – 65 %	25 – 35 %	0 – 7 %	1 – 20 %	< 1 %

Table 5-4 : The percentage range of various Synthesis gas components that can be provided to bioreactors including impurities such as tars and hydrogen sulfide[238,239]

Process outline of the ethanol synthesis gas fermentation method

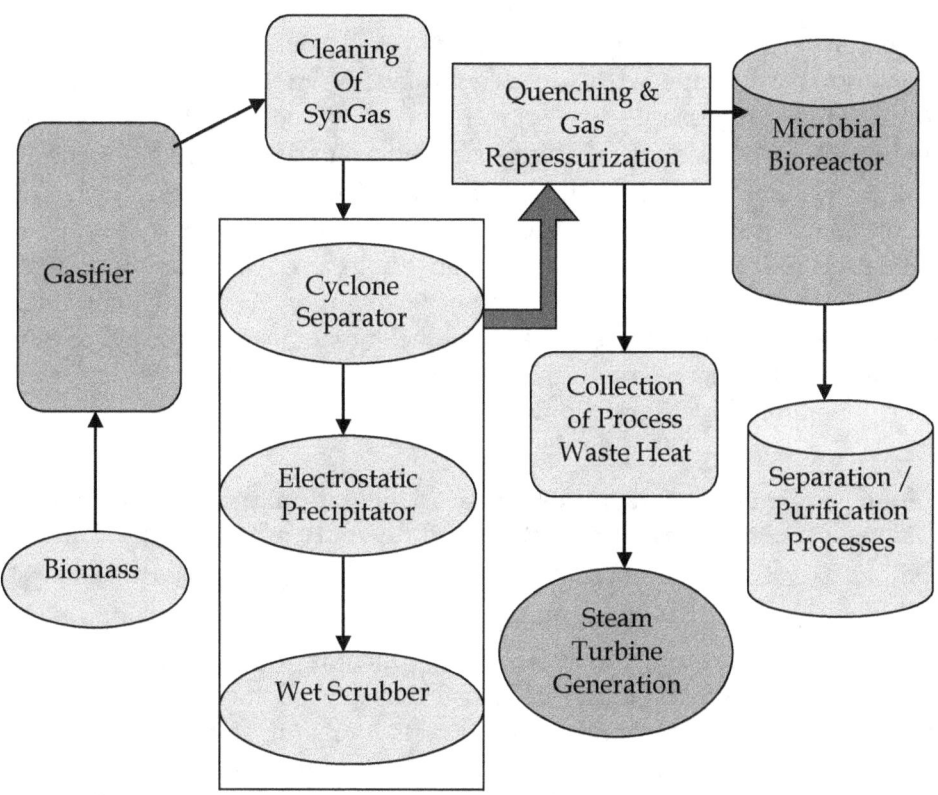

Figure 5-8 : The process description of combined biomass gasification and microbial fermentation of synthesis gas for the production of ethanol

The unit processes of synthesis gas fermentation are outlined in Figure 5-8. The generation of the synthesis gas takes place in the same manner as regular thermochemical gasification. Therefore, the same types of cleaning processes are executed with synthesis gas fermentation manufacture as well. These cleaning steps involve the basic cleaning equipment shown on the left side of Figure 5-8.

This equipment set may include a cyclone separator, electrostatic precipitator and wet scrubber that are applied towards the removal of particulates. After cleaning, the synthesis gas must be cooled down (or quenched) before it is reconditioned (repressurized) and passed into the microbial bioreactor. However, synthesis gas leaving the cleaning process at temperatures above 700 ° C must arrive at the bioreactor at temperatures between 30 – 40 ° C and at pressures slightly higher (1.5 – 2 times) than atmospheric.[237,238] These specific temperature and pressure conditions are necessary for proper microbial assimilation of the synthesis gas.

Therefore, a larger amount of heat must be dissipated with synthesis gas fermentation as compared to conventional thermochemical gasification. Reconditioned synthesis gas in a typical thermochemical gasification refinery arrives at the metal catalytic synthesis stage at around 300 ° C versus the 30 – 40 ° C range that is delivered to the microbial fermenter for synthesis gas fermentation. Thus more heat must be dissipated for the synthesis gas fermentation refinery. However, this means that more waste heat is available for other purposes such as electrical power generation. Currently, companies are now attempting to reprocess the waste heat generated from cooling down synthesis gas and then converting it into steam used towards operating turbine generators for electrical power needs.[240]

5.9 Some advantages of synthesis gas fermentation include lower GHG emissions, efficient energy production and diversity of feedstock utilization

A list of production advantages associated with synthesis gas fermentation is shown in Table 5-5. These include lower carbon footprint, utilizing a wide variety of lignocellulosic based feedstocks and generating electrical power from refinery wastes. In general, alternative ethanol manufacture significantly emits much lower amounts of carbon dioxide. For synthesis gas fermentation, a GHG reduction of at least 80 % or higher is achieved during production when compared to petroleum gasoline manufacture.[241] In addition, synthesis gas fermentation accepts a wider variety of feedstock types, not limited to lignocellulosic types. This includes food/yard, agricultural and industrial wastes. Utilizing a range of feedstock sources allows for a wider choice of placement for refineries to be constructed across the country.

Production capability advantages of the ethanol synthesis gas fermentation method

Carbon footprint	Reduces carbon dioxide emissions by 80 – 90 % (compared to petroleum)
Feedstock sources	Forestry wastes, various industrial wastes, municipal wastes, agricultural residues, energy crops, yard / food wastes, other organic wastes
Refinery power Production	Recovers heat from synthesis gas generation, converts off gas from fermenters into power, reuses spent synthesis gas for power

Table 5-5 : The advantages of lower carbon dioxide emissions (carbon footprint), large array of lignocellulosic or other feedstocks and the various methods of electrical power generation from synthesis gas for the ethanol synthesis gas fermentation method.

In addition, synthesis gas fermentation has the ability to produce electrical power from a number of waste sources. These include waste heat or gaseous based wastes. Gaseous wastes include fermenter off gas or unused synthesis gas. Both types of gaseous wastes are emitted from the bioreactor as either byproduct gases from fermentation or unused portions of carbon monoxide and hydrogen not assimilated by the microbes.

The ability to produce energy from these waste sources results in higher energy efficiency during manufacture at the refinery. This corresponds to an excellent Net Energy Value (NEV) from synthesis gas fermentation when compared to other biofuel manufacturing methods such as corn starch ethanol production. Note that the NEV values from synthesis gas fermentation are comparable to the algae based ethanol values previously discussed in this chapter.

Synthesis gas fermentation also allows for the manufacture of other chemical byproducts during production. This helps to satisfy the integrated biorefinery requirement of chemical based co-products manufacture. There are a variety of chemical byproducts made in addition to the ethanol during the synthesis gas fermentation process. Table 5-6 shows a list of certain anaerobic bacteria that metabolize synthesis gas. These microbes tend to make other chemical byproducts of value besides the ethanol. The majority of these bacterial species are of the genus *Clostridium*. The chemicals produced by these bacteria are ethanol, acetate, butanol and butyrate.

Other than ethanol, acetic acid tends to be the major chemical byproduct made from this type of fermentation. However, other chemicals such as butanol or butyrate can also be made in sizeable quantities. These chemicals have the potential to provide additional product value resulting from the manufacture of other end products. For example, chemical byproducts made from the combination of acetate with other chemicals include ethyl acetate, acetin, vinyl acetate and acetic anhydride.[242] Butanol, as mentioned previously, can be added to gasoline as a fuel additive. Butanol can also produce chemical products of value such as butyl acetate and butyl acrylate.[242]

Specific anaerobic bacteria that produce multiple chemical products through synthesis gas fermentation

Bacterial species	Fermentation products
Bacterium P7	Ethanol, Acetate, Butanol
Clostridium Ljungdahlii	Acetate, Ethanol
Clostridium Carboxidivorans	Ethanol, Acetate, Butryate, Butanol

Table 5-6 : Several synthesis gas fermenting bacteria produce multiple chemical products along with ethanol that also have additional industrial product value[242]

5.10 Several companies now plan on producing ethanol from the thermochemical gasification of municipal waste

Thermochemical gasification based alcohol refineries can produce ethanol or a set of mixed alcohols such as ethanol, propanols, butanols and pentanols from synthesis gas. The production of mixed alcohols from thermochemical gasification is further discussed in the next chapter. In order to selectively produce mixed alcohols or just ethanol from synthesis gas, the temperature and pressure conditions of the synthesis gas must be in a narrower range than it would be for normal hydrocarbon based vehicle fuel using the Fischer-Tropsch process. In addition, the chemical makeup of the catalysts must be modified with materials different than those normally encountered with Fischer-Tropsch catalysts, so that they selectively produce alcohol(s).[243]

Ethanol manufacture from thermal gasification already has a special market niche. Several companies are already planning on making the ethanol in this manner as featured in this section. There are several ways to produce ethanol from thermochemical gasification that takes place in a refinery. One way involves the initial production of methanol followed by subsequent ethanol manufacture. This type of manufacture takes place during the catalytic production stage. After methanol gets produced, it becomes further gasified and then reacts with other synthesis gas components. This mechanism is depicted in Figure 5-9 where methanol, hydrogen and carbon monoxide react on metal catalytic surfaces and then combine together to form ethanol.[244] Afterwards, ethanol vaporizes off the metal catalysts to become condensed and collected as the product. This 'ethanol from methanol' production system has been shown to take place with the NREL mixed alcohol production model and the Enerkem ethanol production method discussed below.

Conversion of methanol to ethanol through catalytically aided reactions

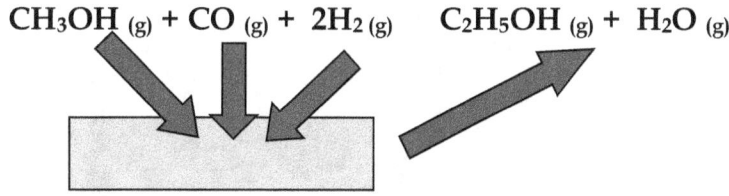

$$CH_3OH_{(g)} + CO_{(g)} + 2H_{2(g)} \quad C_2H_5OH_{(g)} + H_2O_{(g)}$$

Figure 5-9 : Methanol can be converted into ethanol through chemically based mechanisms that form on catalytic surfaces using reactants (including methanol & synthesis gas) in the vapor phase.[244,245]

The company Enerkem happens to be one of the recipients of the 2009 DOE based integrated biorefinery demonstration grants. They obtained a 50 million dollar grant, making them one of the higher funded award recipients. In addition, they are planning to build two other pilot plants in Canada. A valid reason behind funding pilot plants such as theirs has to do with the fact that they can convert municipal wastes, like those contained in landfills, into ethanol. Enerkem's production model plans on producing both ethanol and specialty chemicals obtained from the conversion of methanol.[246] These specialty chemicals are made from further reactions that take place with the produced methanol.

Utilizing municipal waste towards ethanol production is a great idea for biofuel production in general. This practice would assist with environmental remediation efforts such as landfill diversion programs. Municipal wastes applicable towards thermal gasification conversion into ethanol include sources such as organic or food waste, paper and cardboard, cellulose and wood, yard/green waste, textiles and plastics. Non-combustibles materials like metals, glass, construction wastes and electronic wastes cannot be utilized towards the manufacture of synthesis gas. However, other technologies being developed are allowing all municipal waste to be processed together in a refinery setting. One method involves the application of a high energy beam that converts certain waste materials into liquefied solutions valuable enough to be recycled. This allows the processing of materials such as construction waste, glass and metal to be done at a refinery utilizing advanced gasification technology.

Plasma Enhanced Melter (PEM) gasification system effective for converting unsorted municipal waste into synthesis gas

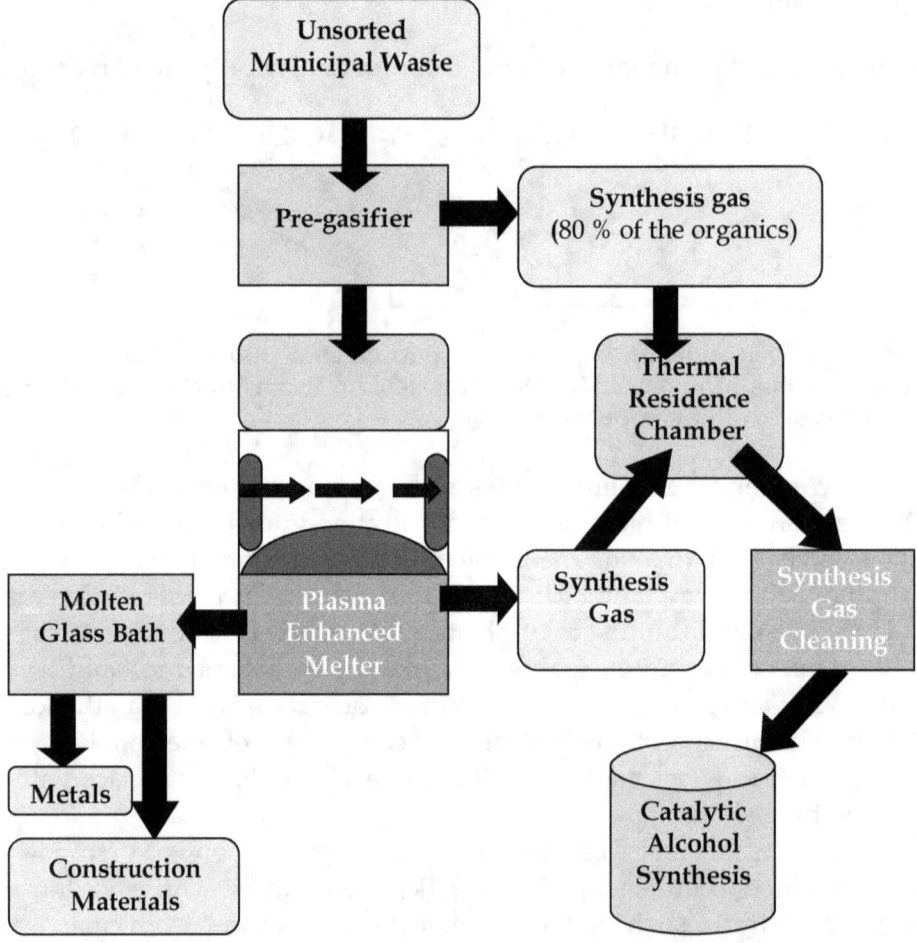

Figure 5-10 : A representation of the plasma enhanced melter (PEM) system that effectively processes all types of (unsorted) municipal waste that can either be converted into alcohol (ethanol) through synthesis gas based catalytic conversion or recycled as metals or construction materials.[247]

One such system known as the Plasma Enhanced Melter (PEM) developed by the company InEnTec allows for all types of municipal waste to be converted either into synthesis gas or recycled as construction materials and metals.[247] The company Fulcrum Biofuels is utilizing this type of system in order to manufacture ethanol from synthesis gas using municipal waste. Fulcrum Biofuels is working towards getting a pilot plant operational that can process 90,000 tons of post sorted municipal waste per year to manufacture just over 10 million gallons of ethanol.[248] A summary of the unit processes that make the ethanol and other materials is shown in Figure 5-10.

In the production process, unsorted municipal waste initially passes through a pre-gasifier that converts around 80 % of the original organic based material submitted into synthesis gas.[247] The remaining amount of organics, dirt, carbon, metals, glass, etc then go into the Plasma Enhanced Melter (PEM) where the material basically becomes vitrified or gasified. After this step the vitrified solutions are then removed from the bottom of the reactor. Both the synthesis gas emitted from the pre-gasifier and PEM get mixed in a thermal residence chamber. The chamber allows for the necessary gas related reactions to occur within a given period of time. The synthesis gas is then cleaned and passed on for catalytic conversion into ethanol. So in summary, all types of municipal wastes that include combustible and non-combustible based wastes can be applied towards alcohol (ethanol) manufacture. Alternatively, a PEM system might be applied towards the manufacture of other renewable fuels from synthesis gas that would include renewable diesel.

REFERENCES:

212. GM backs ethanol start up Coskata – Chemical and Engineering News Vol 86 No 3 p.11 [2008] by M. McCoy
213. Coskata | Technology -> Advantages : http://www.coskata.com/process/ - accessed May 2014
214. Ineos | Ineos Businesses -> Ineos Bio -> Company -> FAQs : 'How much bioethanol does the INEOS Bioprocess produce from a tonne of waste' – http://www.ineos.com/ - accessed May 2014
215. Zeachem | Technology -> Advantages : http://www.zeachem.com/technology/advantages.php - accessed May 2014 (no longer available)
216. Algenol | Direct to Ethanol -> CO2 Utilization : http://www.algenol.com/ - accessed May 2014
217. Algenol | Corporate Overview -> About Algenol : http://www.algenol.com/about-algenol/ - accessed May 2014

218. **Enerkem to use sorted waste as feedstock in biorefinery** – http://www1.eere.energy.gov/bioenergy/pdfs/ibr_arra_enerkem.pdf - US DOE EERE [Dec 2012]

219. **Department of Energy Recovery Act Investment in Biomass Technologies** – US Department of Energy – EERE – http://www1.eere.energy.gov/biomass/pdfs/arra_summary_factsheet_web.pdf

220. **Thermochemical ethanol via indirect gasification and mixed alcohol synthesis of lignocellulosic biomass** – NREL (National Renewable Energy Laboratory) – [Apr 2007] by S. Phillips, A. Aden, J. Jechura, D.Dayton

221. **Design case summary : Production of mixed alcohols from municipal solid waste via gasification** – US DOE EERE Biomass Program [Mar 2010] – http://www1.eere.energy.gov/bioenergy/pdfs/msw_design_case.pdf

222. **Simultaneous hetereolactic and acetate fermentation in the marine cyanobacterium Oscillatoria limosa incubated anaerobically in the dark** – Archives of Microbiology Vol 151 pgs 558-564 [1989] by H. Heyer, L. Stal, WE Krumbein

223. **Ethanol synthesis by genetic engineering in cyanobacteria** – Applied and Environmental Microbiology Vol 65 No 2 pgs 523 – 528 [1999] by MD Deng & JR Coleman

224. **Metabolic engineering of cyanobacteria for ethanol production** – Energy and Environmental Science Vol 2 pgs 857 – 864 [2009] by J. Dexter & P. Fu

225. **Life cycle energy and greenhouse gas emissions for an ethanol production process based on blue-green algae** – Environmental Science and Technology Vol 44 pgs 8670 – 8677 [2010] by D. Luo, Z.Hu & DG Choi

226. **Integrated pilot scale biorefinery for producing ethanol from hybrid algae** – US DOE EERE [2009] – http://www1.eere.energy.gov/biomass/pdfs/ibr_arra_algenol.pdf

227. **Algenol | Direct to Ethanol® -> The Technology** – http://www.algenol.com/direct-to-ethanol/direct-to-ethanol/ -- accessed May 2014

228. **The energy balance of corn ethanol : An update** – USDA [2001] by H. Shapouri, J. Duffield, M. Wang

229. **Opportunities for renewable bioenergy using microorganisms** – Biotechnology and Bioengineering Vol 100 No 2 pgs 203 – 212 [2008] by B. Rittman

230. **Breaking the biological barrier to cellulosic ethanol : A joint research agenda** – Office of Energy Efficiency and Renewable Energy [Jun 2006] by J. Houghton, S. Weatherwax, J. Ferrell

231. **Comprehensive Biotechnology Vol 3 – Chapter 36.3 The biology of acetic acid producers** – ed. M. Moo-Young [Oxford : Pergamon Press 1985] pgs 713 – 721 by TK Ghose, A. Bhadra

232. **Fermentation of cellulose to acetic acid by clostridium lentocellum SG6 : induction of sporolation and effect of buffering agent on acetic acid production** – Letters in Applied Microbiology Vol 37 pgs 304 – 308 [2003] by R. Tammali, G. Seenayya, G. Reddy

233. **Production of acetic acid by Clostridium thermocellum in batch and continuous fermentations** – Biotechnology and Bioengineering Vol 28 No 5 pgs 678 – 683 [1986] by K. Sugaya, D. Tuse, JL Jones

234. **Zeachem | Technology -> Technology overview** – http://www.zeachem.com/technology/overview.php - accessed May 2014 (no longer available)

235. **Zeachem | Technology -> Products** – http://www.zeachem.com/technology/products.php -- accessed May 2014 (no longer available)

236. **Zeachem | Technology -> Carbon Analysis** – http://www.zeachem.com/technology/LCA.php -- accessed May 2014 (no longer available)

237. **Ethanol and acetate production from synthesis gas via fermentation process using anaerobic bacterium Clostridium Ljundahlii** – Biochemical Engineering Journal Vol 27 pgs 110 – 119 [2005] by H. Younesi, G. Najafour, AR Mohamed

238. **Biomass derived syngas fermentation into biofuels : Opportunities and challenges** – Bioresource Technology Vol 101 pgs 5013 – 5022 [2010] by PC Munasinghe, SK Khanal

239. **Biological conversion of coal and coal-derived synthesis gas** – Fuel Vol 72 No 12 pgs 1673 – 1678 [1993] by KT Klasson, MD Ackerson, EC Clausen, JL Gaddy

240. **Brienergy | Sustainable Energy Resources** – http://www.brienergy.com – accessed May 2014

241. **Ineos | Ineos Businesses -> Ineos Bio -> Company -> FAQs : What greenhouse gas savings are expected from Ineos bioethanol** – http://www.ineos.com/ -- accessed May 2014

242. **Ethanol and acetate production from synthesis gas via fermentation process using anaerobic bacterium Clostridium Ljundahlii** – Biochemical Engineering Journal Vol 27 pgs 110 – 119 [2005] by H. Younesi, G. Najafour, AR Mohamed

243. **Production of mixed alcohols from bio-syngas over Mo based catalysts** – Chinese Journal of Chemical Physics Vol 24 No 1 pgs 77-84 [2011] by S. Qiu, W. Huang, Y. Xu, L. Liu et al

244. **Hetereogeneous catalytic conversion of dry syngas to ethanol and higher alcohols on Cu based catalysts** – ACS Catalysis Vol 1 No 6 pgs 641 – 656 [2011] by M. Gupta, ML Smith, JJ Spivey

245. Mechanistic aspect of ethanol synthesis from methanol under CO hydrogenation condition on MoSx cluster model catalysts – Journal of Molecular Catalysts : A Chemical Vol 329 No 1-2 pgs 77 – 85 [200x] by YY Chen, X Zhao, XD Wen, XR Shi et al

246. Enerkem | Technology platform -> Process – http://www.enerkem.com/en/technology-platform/process.html -- accessed May 2014

247. InEnTec | PEM® Technology -> Process Details – http://www.inentec.com/pemtm-technology/process-details.html - accessed May 2014

248. Fulcrum Bioenergy taps Fluor for EPC work on 10 MGY Sierra Biofuels Project – Biofuels Digest [Sept 2010] – http://www.biofuelsdigest.com/

Chapter 6

Mixed Alcohol Production

6.1 The Carboxylate Counter-Current Fermentation refinery is a great example of integrated biorefinery technology developed by academia

The previous chapter demonstrated some examples of alternative ethanol refineries that qualified as integrative biorefineries according to the DOE Integrated Biorefinery Demonstration grant program. This chapter focuses on a specific type of integrated biorefinery model called **carboxylate counter-current fermentation**. In addition, mixed alcohols from thermochemical gasification using a specialized production model is discussed. Therefore this chapter expresses the importance of developing refinery methods to make mixed alcohols. Mixed alcohols represent a legitimate alternative to ethanol as a fuel additive.

The carboxylate counter-current fermentation refinery produces mixed alcohols from carboxylic acids such as acetic acid, propionic acid, butyric acid, etc. The process is very similar to the fermentative acetic acid conversion refinery covered in the last chapter. This type of biorefinery also has the potential to produce an array of biochemicals as byproducts during biofuel manufacture. In addition, this biorefinery can produce clean distilled water as another byproduct from a manufacturing process similar to the way the company Algenol does it.

Carboxylate counter-current fermentation is integrative in approach since it utilizes all types of biomass based feedstocks that are available for fermentation purposes. It also recycles the leftover unfermentable components that are then directed towards generating electrical power and hydrogen. Similar to other integrative biorefineries, the carboxylate counter-current fermentation refinery has the ability to utilize all feedstock materials provided to them for energy, fuel and byproducts, wasting little to nothing during the production process.

Details regarding specific methods and equipment utilized by this type of biorefinery model towards mixed alcohols production are available to the public domain as literature articles written by academia. This chapter summarizes how such equipment and processing methods work together at the refinery model. Such examples include 1) the processing of feedstocks through the lime pile pretreatment method, 2) the use of the counter-current fermenters, 3) the operation of reactor units that make intermediate organic compounds and 4) utilizing hydrogenation units that convert these intermediates into mixed alcohols. Carboxylate counter-current fermentation is one of the best documented integrative biorefinery models developed from scratch by academia.

The technological development of this biorefinery model has progressed to the point that vehicle fuel such as green gasoline and other chemical byproducts can be made from the intermediate compounds of ketones or esters. A carboxylate counter-current related refinery technology known as MixAlco® has been patented by the company Terrabon.[249] They have been involved with the carboxylate counter-current fermentation technology since its onset as developed by professors at Texas A & M University. Before we get into the specific details of carboxylate counter-current fermentation, it is important to introduce the importance of producing mixed alcohols for vehicle fuel use.

6.2 Mixed alcohol manufacture normally produces a high concentration of ethanol while the higher molecular weight alcohol portion serves as a better fuel additive

Mixed alcohol manufacture can be produced from two basic processing technologies. Similar to ethanol manufacture, thermochemical gasification produces a mixed set of alcohols. This method can make a suite of alcohols necessary for a certain application. These alcohols are made by adjusting both the synthesis gas delivery conditions and the types of metal catalysts implemented. One specific method of thermochemical gasification discussed in the next section (6.3) produces the alcohols of methanol, isobutanol and propanol. In addition, the NREL (National Renewable Energy Laboratory) mixed alcohol refinery model happens to be another thermal gasification method previously introduced in chapter 3.

A second type of mixed alcohol production technology is the carboxylate counter-current fermentation method previously introduced in the last section. This process was developed to manufacture a mixed set of alcohols or hydrocarbons from carboxylic acids that would function as a drop in fuel or fuel additive for gasoline based vehicles. For the sake of correct nomenclature, carboxylic acids are the same compounds as organic acids, so these terms can be used interchangeably. Many types of carboxylic acids are the initial end products obtained from the fermentation process. Afterwards they get converted into carboxylate salts before further processing takes place.

The potential distribution of alcohols made during manufacture is one of the key factors associated with mixed alcohol refinery design. Currently, the majority of mixed alcohol refineries tend to produce ethanol as the major end product. It is usually made above 50 % of total alcohol concentration. Examination of two specific production models involving thermochemical gasification and carboxylate counter-current fermentation demonstrate this point. The distribution of ethanol and other linear alcohols are shown in Figure 6-1 for the NREL mixed alcohol refinery and the esterification method for carboxylate counter-current fermentation production. For the NREL model, a total of 96 gallons of mixed alcohols are made from each ton of biomass. Approximately 80 gallons of this total amount consists of ethanol while another 16 gallons are made up of higher molecular weight alcohols such as propanol, butanol and pentanol.[250] This equates to over an 80 % ethanol concentration for all alcohols made. For the esterification method of carboxylate counter-current fermentation, it appears that ethanol makes up at least 60 % or more of all alcohols produced. The alcohol distribution profile in the figure shows minor percentage amounts of higher alcohols ranging from propanol to heptanol. Each higher molecular weight alcohol makes up less than 10 % of those contributing to the total amount.[251]

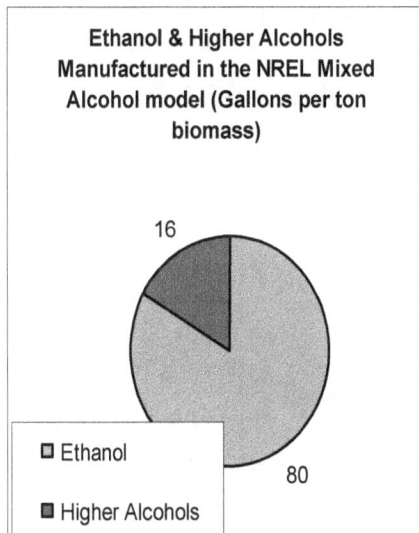
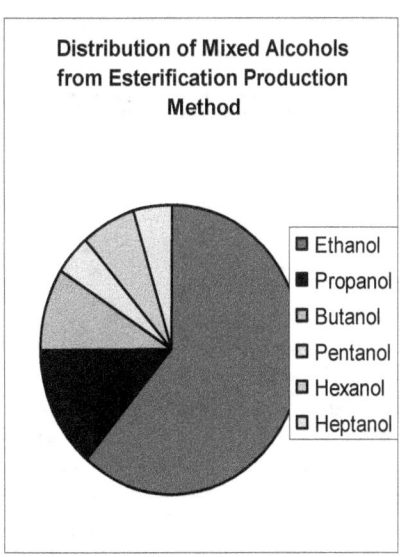

Figure 6-1 : Distribution of ethanol and higher molecular weight alcohols from either the NREL mixed alcohol or carboxylate counter-current fermentation methods where ethanol consists of over 60 % of total alcohol volume for either method.[250,251]

The NREL thermal gasification model involves a much lower percentage distribution of higher molecular weight alcohols made during the manufacturing process when compared to carboxylate counter-current fermentation. This type of result happens for several reasons. First of all, a typical thermochemical alcohol production plant will make a majority of both the alcohols ethanol and methanol under normal circumstances. Secondly, production was geared towards making lower amounts of higher alcohols since they would be sold as solvents in the chemical industry instead of for biofuel distribution.[250]

The separation stage in the NREL refinery helps to establish the alcohol end product distribution as shown in Figure 6-2. It involves a two distillation column setup with recirculation of the methanol.[250] The first column separates the higher molecular weight alcohols from the methanol/ethanol stream. These higher alcohols are then isolated as a product stream that leaves the bottom of the first distillation unit. The methanol/ethanol mixture becomes separated from one another in the second distillation unit. The ethanol gets removed from the bottom of the second distillation column while the methanol goes through a recirculation process. This process gasifies the methanol back into the catalytic production stage.

A two distillation column setup for separating higher alcohols, ethanol and methanol from one another

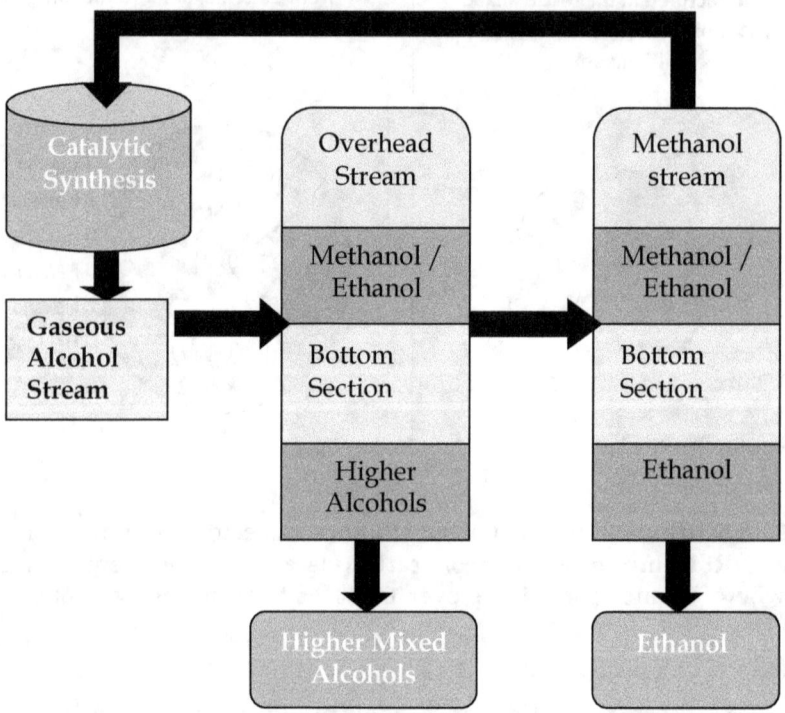

Figure 6-2 : Two distillation columns separate higher mixed alcohols consisting of propanol, butanol and pentanol from an ethanol/methanol stream. The second column then separates ethanol from methanol where the methanol gets recycled back into the catalytic synthesis stage.[250]

The reason why most mixed alcohol refineries produce mainly ethanol has to do with its established purpose as the main gasoline fuel additive. For this reason, much of the research and development associated with mixed alcohol refinery design was centered upon making ethanol as the main product. However, it should be known that producing ethanol as the major component in mixed alcohol manufacture does not have to be the de facto industry standard. For example, many feel that larger molecular weight alcohols would make for a better overall fuel additive rather than ethanol.

There are many reasons why higher molecular weight alcohols would make a better gasoline fuel additive. Larger percentage mixtures of higher molecular weight alcohols impart very good fuel performance behavior such as higher energy value, clean emission gases and high octane ratings. Other important factors that make higher alcohols preferable as fuel additives include favorable chemical/physical properties, better specific fuel consumption (SFC) and good combustion performance. Much of the results obtained concerning the favorable properties of higher molecular weight alcohols have been shown to occur with butanol. Engine performance tests conducted on butanol have demonstrated that this alcohol has improved engine performance metrics such as air to fuel ratio, combustion enthalpy, combustion stability and ignition delay when compared to lower molecular weight alcohols such as ethanol.[252] Indeed, engine performance and wear tests could also be conducted on other higher alcohols or a mixture of them in order to establish their performance in gasoline engines.

Comparison of engine performance issues between high and low molecular weight alcohols

Types of alcohols	Engine performance issues
Methanol, Ethanol	Corrosive, hygroscopic, low SFC values, fuel line vapor lock, good combustion properties, favorable emissions
Propanol, butanol, pentanol, etc	Not as corrosive or hygroscopic, have higher SFC values, excellent combustion properties and favorable emissions

Table 6-1 : Engine performance issues are compared between lower and higher molecular weight alcohols. Both types have good combustion properties and favorable emissions but the lower alcohols also have wear and performance issues due to corrosivity, hygroscopicity and vapor lock as well.

Both methanol and ethanol have fairly good engine combustion properties and assist with decent vehicle emissions. However, methanol and ethanol tend to have the adverse properties of corrosion and hygroscopic (water seeking) nature. Their corrosivity has an adverse effect on engine wear, especially with iron based metals or plastic parts contained in the automobile. In addition, it has been commented that fuel additives such as methanol and ethanol can result in vapor lock of the engine fuel lines due to their high rate of vaporization.[253] A summary as to the comparison of engine performance issues between the lower alcohols of methanol and ethanol versus the higher molecular weight types are shown in Table 6-1.

For the above reasons, a greater percentage of higher molecular weight alcohols should be made in certain mixed alcohol refineries. The concentrations of these alcohols can even be somewhat controlled. For thermal gasification based refineries, it is possible to produce larger amounts of higher alcohols once certain conditions are met. The necessary conditions require specificity in providing the correct pressure, temperature, and synthesis gas consistency as well as providing the appropriate metal catalyst system to be applied towards condensing the alcohols. For example, a mixture of at least 40 % of higher alcohols consisting of propanols, butanols, pentanols and hexanols can be made from catalysts implementing a sol gel type of matrix hosting the metal catalysts.[254]

Carboxylate counter current fermentation can also produce a larger distribution of higher alcohols. For example, the ketonization method of carboxylate counter-current fermentation tends to make higher alcohols when compared to the esterification method. Both these methods are discussed in further sections of this chapter. In addition, slight alterations to the carboxylate counter-current fermentation process could create a majority of higher molecular weight alcohols other than ethanol.

Before getting into the specific details of carboxylate counter-current fermentation it is important to emphasize the potential versatility of thermochemical gasification in making mixed alcohols. In the past it has been possible to 'tailor make' a set of mixed alcohols depending upon the application. This next section talks about such a development that was conducted by the research efforts of academia.

6.3 A certain mixed alcohol thermochemical gasification method was developed in order to help make an ether fuel additive or ready made fuel additive mixture

The thermochemical gasification of biomass can produce a variety of alcohols that include methanol, propanols, isobutanol and also additional hydrocarbons and/or oxygenates. The production of these alcohols and hydrocarbons were done by researchers at the University of Lehigh starting during the late 1980's.[255] There were several purposes for producing a mixed set of alcohols such as these. First of all, the alcohols of methanol and isobutanol can be alternative starting materials used to manufacture MTBE rather than making it from petroleum or natural gas sources. Figure 6-3 below demonstrates how MTBE or ETBE can be produced from the combination of dehydrating the alcohol isobutanol and then reacting isobutylene with either ethanol or methanol during a process called reactive distillation. The second reason was that researchers thought that the combination of such alcohols could be utilized as a ready made fuel additive source. Indeed, the combination of methanol (and sometimes ethanol), isopropanol and isobutanol have already been implemented as a fuel additive in the US and Russia.[253]

The manufacture of these alcohols & hydrocarbons from thermochemical gasification calls for specialized conditions of temperature, pressure and catalyst makeup. Table 6-2 below shows the reaction conditions as well the percentages of various alcohols and hydrocarbons produced during the manufacturing process. These reactions are carried at temperatures between 400 – 450 ° C and at higher pressures (~ 1000 – 1500 psi).[256,257] The types of catalysts used to make the alcohols (and other products) are zinc/chromium materials with the addition of 1 – 3 % potassium or cesium alkali metals.[256,257]

Synthetic production of MTBE or ETBE starting from isobutanol and the use of methanol or ethanol

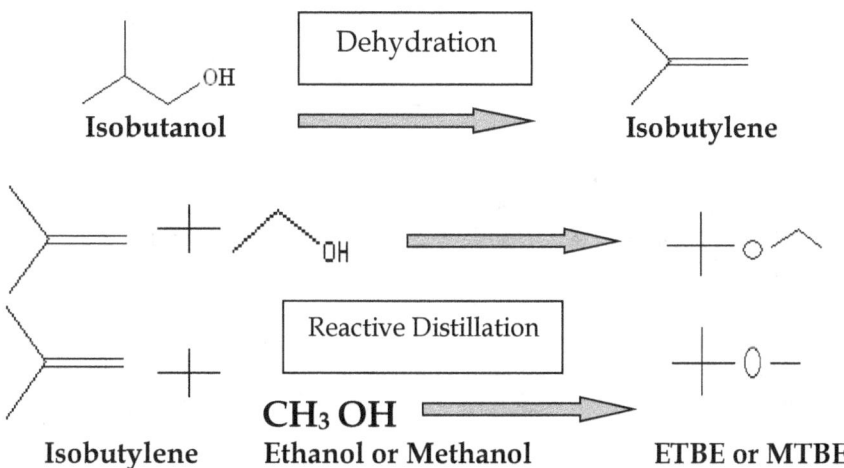

Figure 6-3 : The production of MTBE or ETBE from the dehydration of isobutanol to make isobutylene, which can then be reacted with either ethanol or methanol through the process of reactive distillation to produce MTBE or ETBE.

The numerical values in the table represent the mass production rate at which the alcohols or hydrocarbons are made per kilogram of catalyst material used in the reactor system. The end products include methanol, isobutanol, propanols and oxygenated hydrocarbons. The mass rate amount of each alcohol varies markedly based upon the temperature and pressure conditions along with the amount of alkali metal catalyst additive. At high pressures and lower cesium concentrations it appears that a large amount of alcohols and hydrocarbons are made. The amounts of methanol and isobutanol at 440 ° C, 1500 psi & 1 % cesium are almost twice the amount produced at conditions of 440 ° C, 1000 psi & 3 % cesium. A large concentration of isobutanol produced seems ideal for both ether production and drop in fuel additive purposes.

Production results of alcohols and oxygenates from biomass derived synthesis gas using Zinc/Chromium doped catalysts

	Zn/Cr - 3 % Cs 440 ° C 1000 psig (kg catalyst)*	Zn/Cr - 1 % Cs 440 ° C 1500 psig (kg catalyst)*
Methanol	18 g / kg * hr	32 g / kg * hr
Isobutanol	67 g / kg * hr	115 g / kg * hr
Hydrocarbons	19 g / kg * hr	81 g / kg * hr
Propanols	32 g / kg * hr	8 g / kg * hr

Table 6-2 : Experimental results of methanol, hydrocarbons and higher alcohols produced from synthesis gas using biomass adsorbed onto metal based catalysts mainly consisting of zinc & chromium and doped with cesium or potassium.[256,257,258]
* Rate of production is grams of alcohol(s) or hydrocarbons per kilogram of catalyst per hour

The synthesis gas conditions are also specialized towards producing the various alcohols and hydrocarbons. It is appropriate to have a certain synthesis gas consistency for this production method as is shown in Table 6-3. The carbon monoxide and hydrogen ratio are close to one and both gases are also lower in percentage composition ranging from around 10 - 20 %.[259] The gas composition in general is similar to atmospheric conditions since it may contain at least 50 % of nitrogen content. The gas would also contain some amounts of carbon dioxide as well. Therefore, it appears then that the synthesis gas for this production model is made from thermal gasification conditions that mix normal air with biomass material.

It would be important to further develop thermal gasification production projects similar to this one. As was previously mentioned, this production model was developed to solve a specific problem related to providing a biomass source of MTBE instead of relying upon natural gas or petroleum for it. Even though this application is no longer relevant, it shows the power of utilizing thermochemical gasification to make end products for a specific production purpose. Therefore, specific sets of hydrocarbons or oxygenates can be made from thermochemical gasification by adjusting synthesis gas pressure & temperature conditions in addition to finding suitable catalyst support materials.

Now we know the importance of tailor making a set of mixed alcohols from thermochemical gasification. Just as important is producing mixed alcohols from the well developed method of carboxylate counter-current fermentation. This particular biofuel production method is covered in great detail in this chapter, even more than any other biorefinery mentioned in this book. It is one of the few biorefinery models fully developed by academia and because of this, a lot of the specific details that make for a successful biorefinery are available through the literature articles published in this specific area. It is a biorefinery model that is the ideal blueprint for establishing an integrative biorefinery that has many advantages and important lessons for us to learn concerning biofuel development.

The estimated consistency of synthesis gas produced through an air based thermal gasification method

Synthesis gas conditions	H_2	CO	CO_2	N_2
Air, Biomass, T = 780 – 830 ° C	5 – 16 %	10 – 22 %	9 – 19 %	42 – 62 %

Table 6-3 : The estimated percentages (per volume) of the chemical gaseous components of synthesis gas produced from thermal gasification of biomass and air.[259]

6.4 The carboxylate counter-current fermentation method has many advantages associated with its design

The counter-current carboxylate fermentation platform was developed by Dr. Holtzapple from Texas A & M University with the end result aimed towards producing alcohols or hydrocarbon based biofuel. This is accomplished through a simple fermentation method that is relatively inexpensive and utilizes a variety of lignocellulosic and other waste feedstock sources. The success of this model is associated with the use of common microorganisms plus the ability to utilize feedstock materials that are more readily available. The advantages of this production model make this type of biofuel manufacture worthwhile. These advantages are listed below along with further explanation of some of these points.

- Low cost fermentation due to inexpensive fermenters and naturally occurring microorganisms. There is no need for the addition of enzymes or genetically engineered microbes (like with lignocellulosic ethanol fermentation).
- The counter-current fermentation process can adapt to a wide range of lignocellulosic and waste based feedstocks. This allows for the practical recycling and remediation of waste materials such as municipal waste, manure and sludge.
- The engineering design of the counter-current process allows for widespread recycling and efficient use of materials such as lignins, biosolids and salts.
- The estimated selling price of the alcohols manufactured from the carboxylate counter-current process (MixAlco) 1999 was $0.69 per gallon.[260,261] This price is very competitive compared to other alcohol manufacturing methods. Therefore, due to economic reasons it would be a viable manufacturing process to consider when making alcohols at a large volume production scale.
- Carbon dioxide is recycled through the Pressure Swing Adsorption (PSA) method
- The lime pile pre-treatment method is simpler than and just as effective as the other lignocellulosic pre-treatment methods used to separate lignin from cellulosic material.
- Non-asceptic conditions can be used throughout the production process. Contamination issues are not of a major concern. This also reduces the associated cost of needed equipment.
- Hydrogen generation using thermal gasification can be an effective method to implement the process of hydrogenation common to several biofuel production methods.

Recycling of waste feedstocks assists with remediation : Counter-current fermentation by its very nature allows for a wide diversified use of lignocellulosic or other waste based feedstocks such as manure, sludge, industrial and municipal waste. Recycling material from these types of waste streams has promising potential towards environmental remediation. The carboxylate counter-current fermentation is beneficial for remediation purposes since waste materials like sludge, yard based green wastes, food wastes and municipal wastes such as wood, paper, cardboard, cartons, etc. would normally find their way into landfills but instead are applied towards the production of biofuel.

Chemical equation denoting the removal of phosphates from manure through the process of liming

$Ca(OH)_2$ + HPO_4 $\longrightarrow$ $Ca_5OH(PO_4)_3$
(Lime)   (Phosphate)   (Hydroxyapatite)

Figure 6-4 : Chemical equation (non-stoichiometric) that shows that phosphates can be removed from manure through precipitation into hydroxyapatite.[262]

The recycling of other waste sources such as animal manures also helps to prevent non-source point pollution. Non-point source pollution is where chemicals from fertilizers accumulate in open water bodies and then further contaminate it due to increased microbial or algal growth. In addition, utilizing manure with counter-current fermentation has the potential to recycle phosphorus, which will be a more needed chemical in the future due to supply concerns. The process of removing and recycling the phosphorus can take place through the lime pile pre-treatment method.[262] As shown in Figure 6-4, lime in essence removes or precipitates out phosphate through the formation of a chemical compound called hydroxyapatite that could be removed and isolated from the process.

Balanced nutrient & cellulosic feedstocks: The success of counter-current fermentation has taken place due to the mixture of feedstocks that are available from many sources. A variety of feedstocks allows for the necessary supply of carbohydrates and/or nutrients required for fermentation. A combination of both high carbohydrate and nutrient content materials is essential for the success of carboxylate counter-current fermentation. The combination of these types of biomaterials allows for optimum excretion of carboxylic acids from microorganisms. Nutrient rich feedstocks are favorable towards the natural growth of microorganisms while the carbohydrate rich feedstocks are the preferable source of material for the microbes to feed on. Both types of feedstocks are needed to have an effective counter-current fermentation process. Table 6-4 shows the various types of feedstocks that are either more cellulosic or nutrient based. The suggested mixture of cellulosic to nutrient rich feedstocks in the biomass mixture is 80 : 20.[263]

Mixed cellulosic or nutrient rich feedstocks used in the carboxylate counter-current fermentation process

Cellulosic based feedstocks	Nutrient rich feedstocks
Crop residues, municipal waste, energy crops	Sludge, Manure, Food waste

Table 6-4 : Cellulosic or waste feedstocks are either nutrient rich or high in cellulosic content which are both favorable for counter-current fermentation.

In addition, counter-current fermentation is not limited to the above feedstocks. Other types of renewable sources such as waste glycerol from biodiesel production[264], water hyanciths[265] and processed algae or leftover yeast cells could all be potentially effective towards conducting counter-current fermentation in the future.

The isolation of microbes from natural sources : There may be a need to have a bioconsolidation process of the microbes utilized for conventional lignocellulosic ethanol fermentation. Bioconsolidation denotes that the same microbes can break down cellulose/hemicellulose as well as produce the chemical end products. A bioconsolidation process is already used with counter-current fermentation. The microbes placed into counter-current fermentation reactors are obtained from natural sources such as lake or marine sediments, compost piles or rumen fluid (digestive fluids from cows).[266] The bacteria contained in these natural sources can be isolated easily and grown inexpensively.

For example, microbes can be isolated from a lake sediment sample. After sediment collection, chemicals such as sodium sulfite or cysteine are added to the sample in order to disallow the growth of certain other bacteria in it.[263] These chemicals inhibit oxygen uptake so that anaerobic type bacteria grow instead of aerobic types. In addition, the growth of other natural microorganisms called methanogens has to be inhibited during counter-current fermentation. Methanogens remaining in the fermentation system tend to convert the carboxylic acids into methane gas. In other words, the elimination of methanogens becomes beneficial to counter-current fermentation since having more carboxylic acids available as end products are preferrable to forming methane as a product instead. One of the most effective ways to inhibit the growth of methanogens involves controlling the pH in the nutrient media. A lower pH in the range from 5-6 tends to prevent their growth in the fermenters.[267]

Recycling of the carbon dioxide in the carboxylate counter-current refinery: Pressure swing adsorption (PSA) effectively sequesters carbon dioxide in the carboxylate counter-current refinery. In this system the thermal gasification of biomass produces synthesis gas that contains a sizeable amount of hydrogen. In order to create more hydrogen from synthesis gas, steam is added to it after its generation. This creates hydrogen and carbon dioxide from the carbon monoxide portion of the gas as shown in Figure 6.5. The chemical mechanism responsible for this is known as the water shift reaction discussed earlier in chapter 3 (section 3.3).

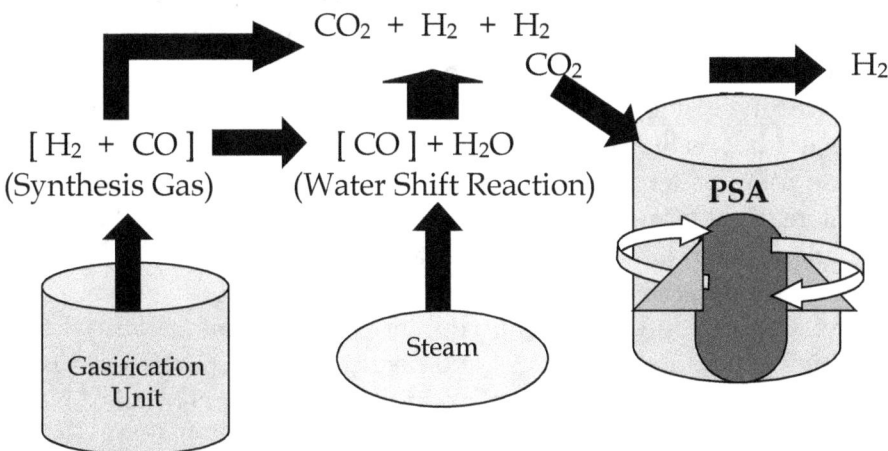

Figure 6-5: Synthesis gas contains carbon dioxide that is separated out by a PSA (pressure swing adsorption) unit.

In the carboxylate counter-current refinery the carbon dioxide becomes sequestered after the additional production of hydrogen takes place. This particular recycling step effectively separates and removes carbon dioxide from a mixture of both gases (hydrogen and carbon dioxide) by utilizing a piece of equipment known as the pressure swing adsorption (PSA) unit as displayed in Figure 6-5. There the carbon dioxide gets extracted out from the stream of both gases through an adsorption and collection mechanism that takes place in the PSA unit. PSA technology has been well developed and implemented effectively with counter-current fermentation. In essence, it is another useful carbon dioxide sequestration technology.

6.5 The counter-current carboxylate mixed alcohol production method is a well developed engineering design with many unique & pertinent unit processes

The research done with countercurrent fermentation is well documented and has been performed as experimental and pilot scale refinery models for sometime now.[260,261] A figurative outline demonstrating the overall carboxylate counter-current fermentation process is shown in Figure 6-6. The main engineering process that produces sufficient quantities of carboxylic acids is counter-current fermentation executed in large fermenters. These fermenters process biomass for long periods of time ranging from 30 – 60 days. They are designed to allow for adequate mixing conditions to take place between the biomass, microbes and nutrient media.

However, before counter-current fermentation can be executed a pretreatment method is necessary to carry out in order to properly prepare the biomass for fermentation. This type of process is known as the lime pile pretreatment method.[261,267] This method works by adding lime to piles of feedstock material. The piles are periodically turned or moved around. This type of treatment allows for the effective separation of lignins from the cellulose and hemicellulose material. The pre-treatment piles are often rotated in a round robin pattern as shown in the figure with holding times similar to the countercurrent fermenters. Therefore, it can take several months to process lignocellulosic material from the time it enters the lime pre-treatment piles until it travels through all of the counter-current fermenters.

After lime pile pre-treatment, counter-current fermentation takes place using mainly a mixture of cellulose and hemicellulose separated out from the lignocellulosic material originally placed in the pre-treatment piles. Naturally occurring microbes transferred to the fermenters help execute the fermentation process. Effective fermentation involves efficient cross mixing between the microbes, nutrients in the media and the biomass material. A typical refinery may have a series of 6 different fermenters as shown in the figure. Cross mixing takes place as nutrient media & microbes introduced from one end fermenter meets up with biomass input at the opposite end fermenter. In other words, the biomass and nutrient media get transferred to consecutive fermenters going in opposite directions as depicted by the arrows in the figure. The rule of thumb is that increasing amounts of carboxylic acids are made as more recent, fresh amounts of biomass [cellulose and hemicellulose] meet the older nutrients and microbes traveling from the opposite end fermenters. Each transfer step (going in the counter-clockwise direction) tends to produce more carboxylic acids resulting from the ideal mixing conditions in the fermenters.

A more detailed process outline of the carboxylate counter-current fermentation method for mixed alcohols or hydrocarbons

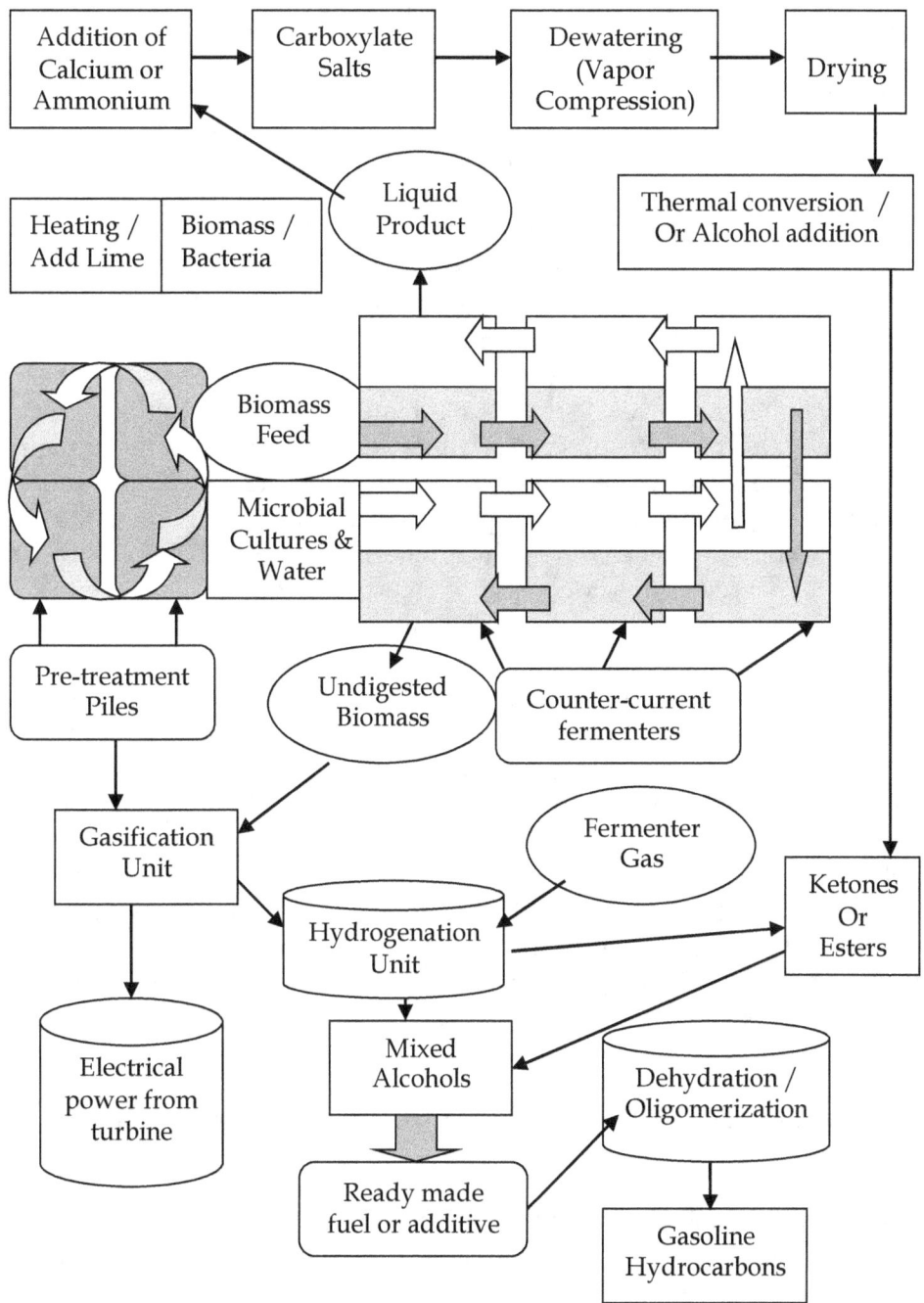

Figure 6-6 : The counter-current carboxylate process uses a series of pre-treatment piles and fermenters that produce carboxylate salts where biomass and microbes/water are fed in opposite directions

On one end, the biomass that did not get metabolized by microbes exits the fermenter units as undigested biosolids. At the opposite end fermenter, a liquid product containing a high concentration of carboxylic acids gets removed for further product processing. These carboxylic acids then become converted into carboxylate salts in another reaction vessel upon the addition of calcium or ammonium bicarbonate. The salts are then dewatered through the use of a method called vapor compression. This treatment implements heat exchangers that produce heated water vapors that help to vaporize the water from the fermentation broth.[261] The vaporized water can then be gathered and collected as distilled water. Afterwards, the carboxylate salts are then concentrated, further dried and crystallized.

The intermediate products of ketones or esters are formed from the carboxylate salts in a reaction vessel involving special thermochemical conditions. The ketonization process utilizes the addition of salts and heat to make the compounds while esterification requires alcohol addition and heat. Usually just one of these processes, either ketonization or esterification, is executed, depending upon the configuration of the particular refinery. Afterwards, a final thermochemical process turns these compounds into a set of mixed alcohols. This involves the process of hydrogenation. Hydrogen added to a particular reactor vessel converts ketones or esters into their respective mixed alcohol structures. The mixed alcohols can then removed as a ready made drop in fuel or additive. Alternatively they can be further refined into a set of fuel grade mixed hydrocarbons by using a dehydration and isomerization unit. This unit operates at a relatively high temperature ($\sim$ 300 °C) with the use of zeolite type catalysts.[263]

In another location of the refinery, thermochemical gasification of waste materials assists towards the generation of hydrogen as well as the providence of electrical power. To accomplish this, the thermochemical gasification process utilizes the waste products of lignins from pretreatment and undigested biosolids from fermentation. In addition, another source of hydrogen for the refinery is obtained from the waste gas emitted from the large fermenters. This fermenter gas consists of a combination of hydrogen and carbon dioxide.

6.6 Many chemical reactions are responsible for determining an array of mixed alcohol products

Many chemical reactions take place in order to form intermediate products and mixed alcohols with carboxylate counter-current fermentation. This section reviews the pertinent reactions that make these intermediate compounds as well as the end product distribution of mixed alcohols or hydrocarbons. Although not an extensive amount of chemistry is presented in this book, the author thought it important to include these organic chemistry based reactions that characterize the overall counter-current fermentation production process. This information should be relevant to those who have a deep interest in biofuel production.

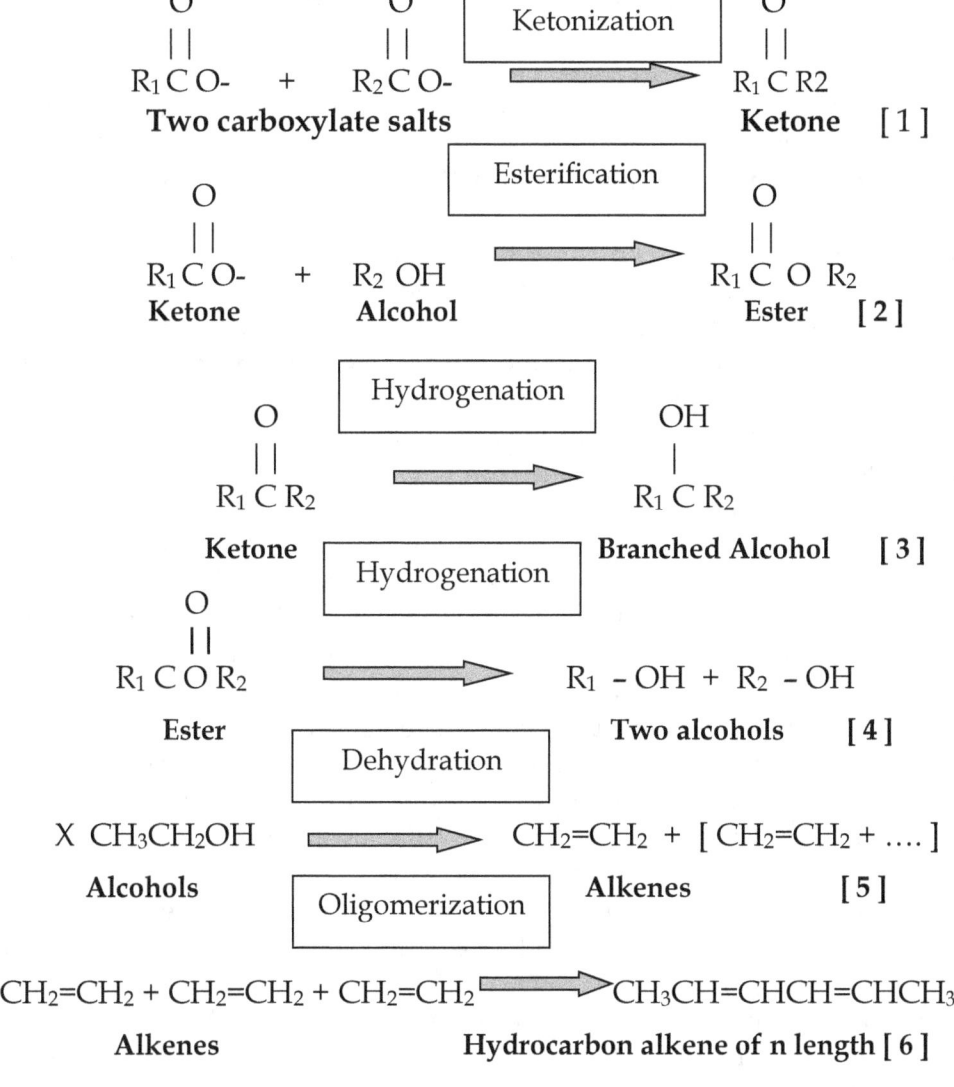

Figure 6-7 : The basic reactions that take place during the carboxylate counter-current fermentation process include esterification and ketonization as well as hydrogenation. Hydrogentation of an ester produces two separate alcohol molecules. Dehydration and oligomerization also take place to form hydrocarbons.

The basic chemical reactions that take place with the carboxylate salts and intermediate products are shown in Figure 6-7. A carboxylate salt consists of a backbone of hydrocarbons where $R_x = C_nH_{2n+1}$. For example, acetate is written as R_1 (C=O) O- where R_1 = CH_3, therefore the structure is actually CH_3 (C=O) O-. For propionate R_1 (C=O) O- where R_1 = CH_3CH_2- and the structure is then CH_3CH_2 (C=O) O-.

The first reaction covered in the figure for the carboxylate counter-current process is ketonization. The reaction involves two different carboxylate salts combining together under high thermal conditions [1]. A ketone compound is the resultant intermediate product formed. Another alternate reaction that takes place with a carboxylate salt is its combination with an alcohol which results in the formation of an ester compound [2].

Both of these compounds then undergo a hydrogenation reaction where hydrogen gets mixed in with either the ketone or ester. A branched alcohol becomes formed in the case of ketone conversion [3] whereas two separate alcohols are produced during the hydrogenation of an ester [4]. The last set of reactions relate towards converting alcohols into hydrocarbons where the hydrocarbons can be utilized as a separate stand alone fuel. To make these hydrocarbons the mixed alcohols are dehydrated [5] to form alkenes. These alkenes then go through an oligomerization process [6] where they combine together to form larger sized hydrocarbons of various lengths.

The specific carboxylic acids emitted during fermentation determine the types of alcohols produced later on during thermochemical processing. The original carboxylic acids contained in the product media direct the yield and distribution of the respective alcohols made. This takes place because there is variation in the percentage composition of the specific types of carboxylic acids formed during carboxylate counter-current fermentation.

Table 6-4 shows a list of the carboxylic acids and their corresponding alcohols made from the esterification thermochemical process. For alcohols made from esterification, the carbon size of the alcohol product is equivalent to the number of carbons contained in the carboxylic acid it was made from. These relationships are expressed in the table. For example, the four carbon carboxylic acid named butyric acid makes the four carbon alcohol called butanol.

Figure 6-8 shows some examples of the possible ketones and resulting alcohols that are formed from the ketonization and hydrogenation processes. For example, the combination of acetate and propionate makes the ketone compound of methyl ethyl ketone. Upon further hydrogenation this ketone turns into 2-butanol. The same processes apply for the production of isopropanol and 2-pentanol made from a set of carboxylate salts.

Carboxylic acids corresponding to linear alcohol products made esterification

Carboxylic Acid	Corresponding Alcohol
Acetic Acid (2 carbons)	Ethanol (2 carbons)
Propionic Acid (3 carbons)	Propanol (3 carbons)
Butyric Acid (4 carbons)	Butanol (4 carbons)
Valeric Acid (5 carbons)	Pentanol (5 carbons)
Caproic Acid (6 carbons)	Hexanol (6 carbons)

Table 6-4 : A list of carboxylic acids and the corresponding straight chain alcohols that are produced through the hydrogenation of esters from carboxylate salts.

Formation of brached alcohols from ketonization by combining acetate with other types of carboxylates

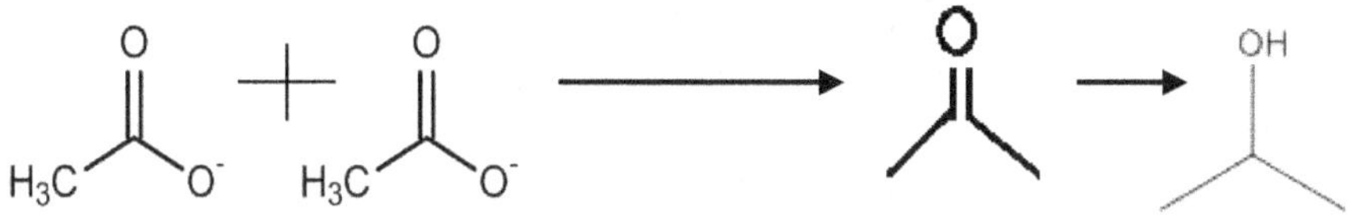

Acetate + Acetate → Acetone → Isopropanol

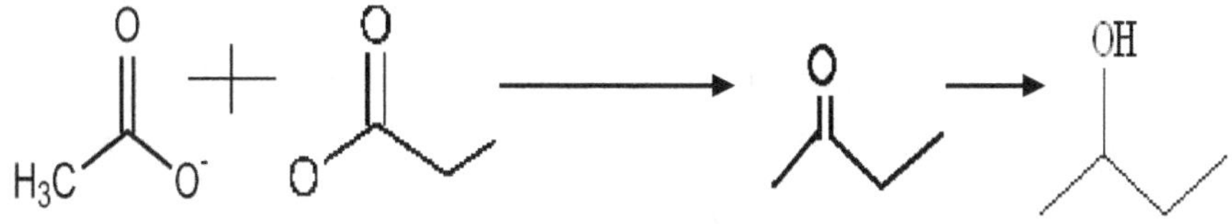

Acetate + Propionate → Methyl ethyl ketone → 2-Butanol

Acetate + Butyrate → Methyl propyl ketone → 2-Pentanol

Figure 6-8: The production of branched alcohols from ketonization using the combination of acetate and other carboxylate salts. The processes above omits the steps of formation of calcium carboxylate salt, ketonization and hydrogenation.

An array of 3-9 carbon branched alcohols made from the combination of 2-5 carbon carboxylate salts from ketonization

	Acetate	Propionate	Butyrate	Valerate
Acetate	2-Propanol	2-Butanol	2-Pentanol	2-Hexanol
Propionate	2-Butanol	3-Pentanol	3-Hexanol	3-Heptanol
Butyrate	2-Pentanol	3-Hexanol	3-Heptanol	4-Octanol
Valerate	2-Hexanol	3-Heptanol	4-Octanol	5-Nonanol

Table 6-5 : A 4 x 4 matrix of branched alcohols made from the combination of two carboxylate salts from 2-5 carbons in length. A total of 9 different alcohols are possible from these combinations.

When carboxylate salts, having from two to five carbons, are mixed together, nine possible branched alcohols can be formed from the ketonization process. This is demonstrated by the 4 x 4 matrix shown in Table 6-5. The combination of acetate through valerate on the top row mixed with one of the same types of carboxylic acids listed in the first column produces a branched alcohol corresponding to the appropriate cell in the matrix. These branched alcohols range in size from 3 carbons (i.e., 2-propanol) to 9 carbons (5-nonanol). The length of the alcohol formed follows the formula (n + m) − 1 where n & m are the carbon lengths of the perspective carboxylic acids combined together.

6.7 Recycling of alcohols and carbonate take place during esterification and ketonization processes

Figure 6-8 part 1 shows that half the alcohols made during the esterification process are further recycled again to form more esters. This takes place since two alcohols are formed during the hydrogenation process. Hydrogenation basically breaks apart an ester compound into two alcohols upon exposure to hydrogen. One of these alcohols then gets recycled and recombines with a carboxylate salt to form an ester during the esterification process. An example of this process, as outlined in Figure 6-8, shows that the two alcohols of ethanol and butanol are made after hydrogenation happens. The ethanol then recirculates back into the esterification production stage to form the ester ethyl butyrate

The compound calcium carbonate also gets recycled upon its application towards producing the corresponding ketones during the ketonization process as shown in Figure 6-8 part 2. Thermochemical processing breaks apart calcium carboxylate into the ketone acetone and calcium carbonate. The calcium carbonate then gets applied back into the original ketonization process where it combines with two acetate compounds to form the calcium carboxylate salt again.

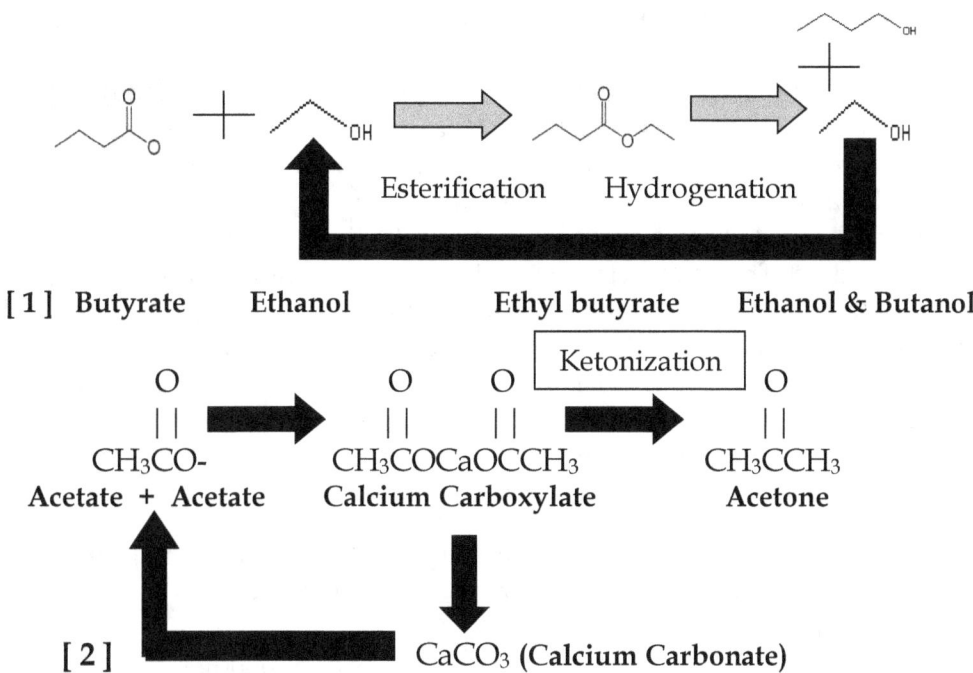

Figure 6-8 : Alcohols and carbonates are recycled during the esterification and ketonization processes of carboxylate counter-current fermentation.

The reuse of carbonates and alcohols demonstrated in Figure 6-8 are two ways materials are recycled to enhance further mixed alcohol production through the counter-current fermentation process. Materials other than calcium carbonate and alcohols are also recycled to form further power and hydrogen generation. This takes place with the waste biosolids leftover after fermentation and the lignins separated from the lime pile pretreatment method as was discussed in section 6.4.

6.8 Certain cultivation and process conditions can produce a large favorable distribution of high molecular weight alcohols

It should be preferable to produce a larger percentage distribution of higher molecular weight carboxylic acids during carboxylate counter-current fermentation. A distribution of larger sized alcohols would be made from these higher molecular weight carboxylic acids. A mixture of mostly higher molecular weight alcohols would make a better fuel additive. There are several ways in which large molecular weight alcohols can be produced due to fermentation and further processing conditions.

Feedstock types in large part are the main factor that determines carboxylic acid distribution in the resulting product fermentation broth. A larger distribution of higher molecular weight of carboxylic acids tends to be made from feedstock sources such as fats and proteins. Fermentation of fats or proteins produces lesser amounts of acetic acid and higher amounts of larger carboxylic acids. It has been shown that protein hydrolysis and resulting amino acid metabolism from such bacteria tend to produce propionic and butyric acids.[268] This is the case for feedstocks such as food wastes, algae and yeast material that contain higher amounts of fats and proteins and will result in producing larger sized carboxylic acids as is summarized in Figure 6-9 below. One source of potential material that can be input into carboxylate counter-current refineries is the waste residual biomass obtained from other types of biofuel refineries. For example, leftover algae and yeast cells derived from the fermenter units of other refineries could be utilized for this purpose.

The recycling of certain waste residues leading towards the production of higher molecular weight carboxylic acids

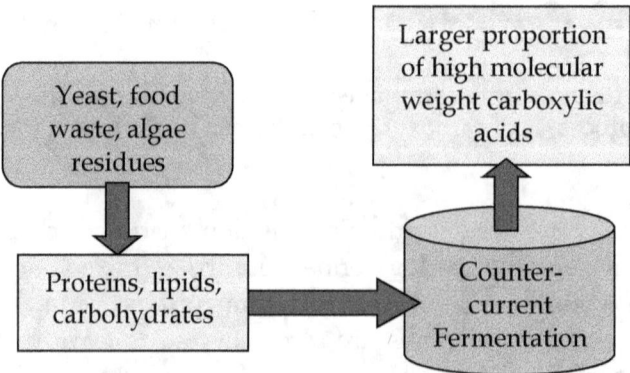

Figure 6-9 : Residues such as spent yeast, microbe or algae cells from other biorefineries could assist in helping to manufacture higher molecular weight carboxylic acids due to the high amounts of lipids and proteins they contain.

In addition, there may be other manufacturing methods that allow for the production of higher molecular weight alcohols preferably over lower ones such as ethanol. This can be accomplished by removing an initial intermediate product before it is further converted into an alcohol. This type of arrangement is summarized in Figure 6-10. An example of this is the removal of the lowest molecular weight ester ethyl acetate from a mixture of other esters after the esterification of the carboxylic acids has taken place. The removal of ethyl acetate allows for its sale as a valued chemical byproduct. In addition, it reduces the amount of ethanol produced in the refinery.

The possible removal of ethyl acetate from the esterification based counter-current fermentation process

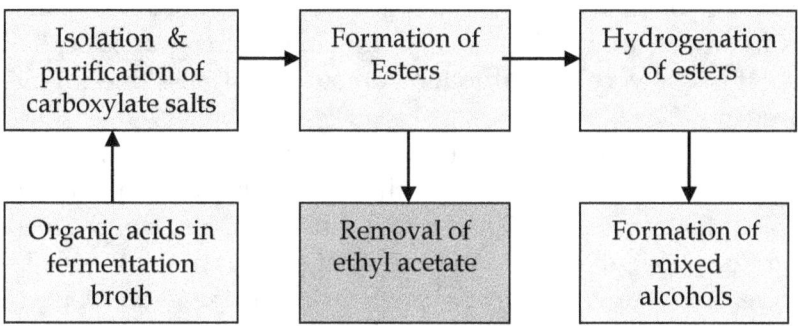

Figure 6-10 : Larger amounts of higher molecular weight alcohols could be produced if the formation of ethanol was partially substituted by the process of removing ethyl acetate from the mix of esters made after the esterification process.

Respective boiling points of certain ester compounds

Ester compounds	Boiling points
Ethyl Acetate [2 carbons]	77 °C
Ethyl Propionate [3 carbons]	98.9 °C
Ethyl Butyrate [4 carbons]	121 °C

Table 6-6 : The different boiling points of the lower molecular weight esters formed from the esterification process of carboxylate counter-current fermentation

Table 6-6 below provides the various boiling points of ethyl acetate and other higher molecular weight esters. Utilizing this information, the removal of only ethyl acetate from a mixture of esters could be implemented if the solution is heated just below 100 ° C. This allows the ethyl acetate to be removed through distillation while all the other esters remain in solution and travel on to the next hydrogenation stage.

6.9 Effective hydrogen gas generation helps to produce reasonably priced mixed alcohols

Hydrogen gas can be manufactured through the joint processes of synthesis gas generation and steam input as previously discussed in this chapter. This type of hydrogen production has been extensively outlined by the Department of Energy (NREL department) as a production model that generates hydrogen from a biorefinery utilizing biomass specifically for this purpose.[269] It is also very similar to how other current biorefineries make their own hydrogen.

According to this model, NREL estimated that the minimum selling price of hydrogen per kg would be $1.38 if the cost of the feedstock were $30 per ton for a plant that processes 2000 tons per day.[269] The price of the feedstock is the major component that determines the overall cost of hydrogen manufacture. It is estimated that the feedstock contributes to around 30 % of total production cost. Other unit processes that affect hydrogen cost of production include the handling and drying of biomass, the gasification and cleaning process, gas compression and steam reforming – pressure swing adsportion.

Research concerning hydrogen manufacture within the carboxylate counter-current refinery has projected overall mixed alcohol production costs as functions of feedstock and hydrogen production costs as is shown in Figure 6-11. It can be ascertained from the figure that mixed alcohol production costs rise in accordance with higher hydrogen production costs.[260] The overall price in mixed alcohol production is directly affected by the range of hydrogen production costs obtained from the waste biomass sources of the refinery. This hydrogen production cost is measured as the dollar amount per kilogram of hydrogen made. So, it appears that the range of this cost can vary from $1 to $3 per kilogram of hydrogen made. These results are also correlated to how much the feedstock material itself costs. A good case scenario for reasonable hydrogen cost from a typical carboxylate counter-current refinery would be priced around $1 per kg.[270] Therefore, if the feedstock prices themselves were also reasonable in purchase price at around $60 per ton or less, then the refinery would be able to manufacture alcohols at a very competitive cost of around $0.80 - $1.00 per gallon.[260]

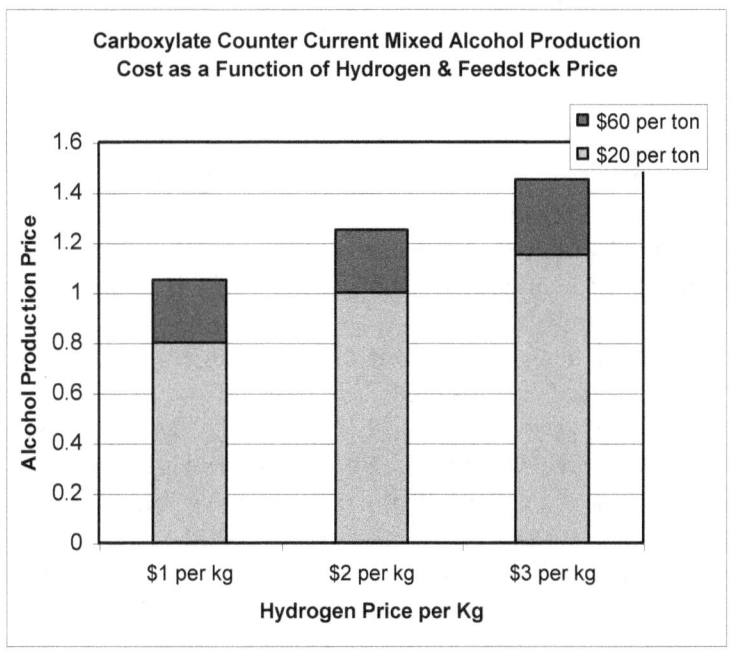

Figure 6-11 : The projected carboxylate counter-current fermentative alcohol production costs as a function of hydrogen price and feedstock cost. Figures for graph estimated from several other charts.[260]

REFERENCES:

249. Terrabon → MixAlco® Overview - http://www.terrabon.com/mixalco_overview.html - accessed August 2014
250. **Thermochemical ethanol via indirect gasification and mixed alcohol synthesis of lignocellulosic biomass** – NREL (National Renewable Energy Laboratory) – [Apr 2007] by S. Phillips, A. Aden, J. Jechura, D.Dayton
251. **Producing mixed alcohols from biomass** – by M. Behring, J. Swenton [2007] – http://www.ou.edu/class/che-design/a-design/projects-2007/Mixalcs-Presentation.pdf - accessed May 2014

252. **Evaluation of butanol-gasoline blends in a portable fuel injection, spark ignition engine** – Oil & Gas Science and Technology Vol 65 No 2 pgs 345-351 [2010] by J. Dernotte, C. Mounaim-Rouselle, F. Halter, D.Seers

253. **Increase of the environmental and operational characteristics of automobile gasolines with the introduction of oxygenates** – Theoretical Foundations of Chemical Engineering Vol 43 No 4 pgs 563 – 567 [2009] by AV Tsarev, SA Karpov

254. **Preparation of K-Co-Mo-C catalyst and its catalytic performance for mixed alcohol synthesis** – Chinese Journal of Catalysis Vol 27 pgs 409 – 415 [2006] by CL Li, YL He, ZM Liu

255. **Higher alcohol and oxygenate synthesis over cesium doped Cu/ZnO catalysts** – Journal of Catalysis Vol 116 pgs 195 – 221 [1989] by JG Nunan, CE Bogdan, K. Klier, KJ Smith et al

256. **Reaction and surface characterization study of higher alcohol synthesis catalysts : I. K-Promoted commercial Zn/Cr spinel** – Journal of Catalysis Vol 169 pgs 438 – 446 [1997] by WS Epling, GB Hoflund, WM Hart, DM Minahan

257. **Reaction and surface characterization study of higher alcohol synthesis catalysts: II. Cs-Promoted commercial Zn/Cr spinel** – Journal of Catalysis Vol 172 pgs 13 – 23 [1997] by WS Epling, GB Hoflund, WM Hart, DM Minahan

258. **Advances in catalytic synthesis and utilization of higher alcohols** – Catalysis Today Vol 55 No 3 pgs 233 – 245 [2000] by RG Herman

259. **Hot gas removal of tars, ammonia and hydrogen sulfide from biomass gasification gas** – Catalysis Reviews Vol 49 pgs 407 – 456 [2007] by W. Torres, SS Pansare, JG Goodwin Jr

260. **Biomass conversion to mixed alcohol fuels using the MixAlco process** – Applied Biochemistry and Biotechnology Vol 77-9 pgs 609-631 [1999] by MT Holtzapple, RR Davidson, MK Ross et al

261. **Techno-economic analysis of biomass to fuel conversion via the MixAlco process** – Journal of Industrial Microbiology and Biotechnology Vol 37 No 11 pgs 1157-1168 [2010] by V.Pham, M.Holtzapple, M. El-Halwagi

262. **Production of renewable bioproducts and reduction of phosphate pollution through the lime pretreatment and acidogenic digestion of dairy manure** – Environmental Progress and Sustainability Energy Vol 28 No 1 pgs 121-127 [2008] by ED Blackman, GP van Walsum

263. **Investigation of nutrient feeding strategies in a counter-current mixed acid multi-staged fermentation : Experimental Data** – Applied Biochemistry and Biotechnology Vol 16 no 4 pgs 426-442 by AD Smith, NA Lockman, MT Holtzapple

264. **Effect of biodiesel glycerol type and fermentor configuration on mixed acid fermentations** – Bioresource Technology Vol 101 No 23 pgs 9185 – 9189 [2010] by AK Forest, R. Sierra, MT Holtzapple

265. **Effects of temperature and pre-treatment conditions on mixed acid fermentation of water hyanciths using mixed culture of thermophillic microorganisms** – Bioresource Technology Vol 101 No 19 pgs 7510 – 7515 [2010] by AK Forest, J Hernandez, MT Holtzapple

266. **Conversion of municipal solid wastes to carboxylic acids by thermophillic fermentation** – Applied Biochemistry and Biotechnology Vol 111 No 2 pgs 93-112 by WN Chan, MT Holtzapple

267. **Carboxylate Platform : The MixAlco Process Part 2 – Process Economics** – Applied Biochemistry and Biotechnology Vol 156 pgs 537-554 [2009] by CB Granada, MT Holtzapple, G.Luce et al

268. **Waste to bioproduct conversion with undefined mixed cultures : the carboxylate platform** – Trends in Biotechnology Vol 29 No 2 pgs 70 – 78 [2011] by MT Agler, BA Wrenn, SH Zinder et al

269. **An overview of aqueous-phase catalytic processes for production of hydrogen and alkanes in a biorefinery** – Catalysis Today Vol 111 pgs 119-132 [2006] by GW Huber, JA Dumesic

270. **Biomass to hydrogen production detailed design and economics utilizing Batelle Columbus Laboratory indirectly heated gasifier** – National Renewable Energy Laboratory [2005] by P.Spath, A. Aden, T. Eggeman, M.Ringer, B. Wallace

Chapter 7

Biofuels made from Algae

7.1 Algae produce many types of biofuels in a number of different ways

Algae are one of the most flexible biomass sources that can be utilized for biofuel production. Algae can not only produce all types of biofuels but they are able to make them in a number of different ways. In addition, algae have the potential to make several different types of biofuels at one time. Some examples of the aforementioned points include the following: 1) Similar to lignocellulosic material, algae can produce a bio-oil that has the potential to be fashioned into many vehicle fuels. 2) Algae that go through fermentation are known to emit hydrogen in addition to the alcohol they produce. 3) Genetic engineering may allow algae to make several biofuels at one time through the incorporation of a number of metabolic pathways introduced into the algae.

The manner that the biofuel is produced from algae depends mostly on how they are originally cultivated. For example, high protein algae are usually cultivated under normal conditions whereupon afterwards they pass through some low thermochemical processes to produce both animal food and renewable fuel. Algae that accumulate a high concentration of lipids require alteration of growth conditions and are further processed to extract and convert these lipids into renewable diesel or biodiesel.

Algae also contain saccharide sources that in the future may allow for several choices regarding further biofuel production. Saccharides as a feedstock source from algae can be utilized towards either fermentative or thermochemical biofuel production. In general, regardless of the source of saccharides, the thermochemical production method of choice may be the aqueous phase processing of saccharide units such as glucose. A wide variety of biofuel compounds have been known to originate from glucose after thermochemical processing has been executed. Production technologies such as aqueous phase processing should become prevalent as more accessible forms of saccharides obtained from algae (or other biomass) are available. One preferred mode of saccharide production includes excretion of the saccharides from within the algae.

In fact, the excretion of potential biofuel compounds from algae appears to be the more modern and upcoming way to produce biofuel in general. This includes the manufacture of alcohols such as ethanol. Hydrocarbon compounds can also be emitted from algae. These types of compounds include alkanes, isoprenoids, free fatty acids, aldehydes and fatty alcohols.

This 'direct' method of biofuel production usually requires genetic modification of the algae. Genetic engineering of algae allows the proper enzymatic systems to be put into place in order to produce end products in appreciable quantities. There are also circumstances where these compounds can be made from natural algae. They also require specialized cultivation conditions. However, even though normal, regular enzymes in natural algae produce the same compounds as genetically altered microorganisms, they oftentimes do it in much lower micro-level concentrations. Indeed, there are challenges that need to be overcome before these systems can be applied to larger scale production, regardless of whether they come from genetically engineered or natural (wild type) algae. These challenges include giving high enough yields and production rates along with developing extraction systems that allow for efficient removal of the compounds from solution.

The main advantage of utilizing algae for direct biofuel production through excretion has to do with the ability of the algae to assimilate carbon dioxide instead of saccharides as their carbon food source. Carbon dioxide is a renewable resource that will become more available due to improvements in manufacturing methods that allow it to be captured or sequestered. There are a number of industries that emit carbon dioxide during manufacture, some of which include electrical energy generation and vehicle fuel production itself. This chapter discusses an example of algae cultivation taking place through carbon dioxide capture. High density algae cultures grow by assimilating the captured flue gases given off by a coal power plant.

This chapter in general gives several examples of biofuel production from the cultivation of algae. Many of these methods are derived from scientific research at universities or company related technological developments. Therefore, this material has a lot of technical information that will take some time and effort to digest. It is meant to spark deeper interest in biofuels production from algae. Even though this chapter covers several topics there are still many more subjects of algae biofuel production that must be further shared by other researchers. Hopefully this subject matter will motivate the reader to find out more about them.

7.2 Indirect ethanol production from algae oftentimes requires an involved pretreatment process

There are two general approaches towards producing alcohols from algae. One method produces alcohols directly while the other does it indirectly. The direct method cultivates algae so that alcohol(s) are the main products made by them. The algae excrete these alcohols into the cultivation solution in appreciable quantities. Much of the material in this chapter deals with alcohols like this that are directly excreted by the algae. This appears to be the more modern and upcoming way to produce alcohols from algae.

On the other hand, the indirect method works on the premise that algae grow saccharides within them initially. These saccharides then get harvested and afterwards are fed to other microbes that ferment them into alcohol. Although several different types of alcohols can be made in this manner, this section covers ethanol production obtained from the fermentation of algae based polysaccharides like starch. The majority of alcohols other than ethanol made through conventional fermentation (of the algae saccharides) usually require genetically modified microbes to produce them. In other words, the algae just provide the saccharide sources for fermentation while genetically engineered microbes are required to convert these saccharides into alcohols other than ethanol.

Conventional fermentation usually takes place with lignocellulosic material containing saccharides from cellulose and sometimes hemicellulose. However, algae can also provide a readily available source of saccharides utilized towards fermentation. For algae, these usually come in the form of starch grains contained within them. Under certain growth conditions, many types of algae are known to accumulate these starch grains. Oftentimes this requires several cultivation steps where the nutrient media lacks specific nutrients. These starch grains can gather up to 60 % of the cell dry weight utilizing a growth method such as this.[271]

After cultivation, the processing steps of pretreatment and fermentation are carried out in order to produce the ethanol. An example of this is the experimental procedure shown in Figure 7-1. The pretreatment process involves the separation and modification of the polysaccharide material from the algae before fermentation takes place. These steps are quite complicated in application. Needless to say, there is a need to develop less complicated and more effective pretreatment methods for saccharide providence from algae. Pretreatment is essential to carry out before fermentation is executed. The pretreatment technique shown in Figure 7-1 requires a two step process involving the methods of liquefaction and saccharification. This particular liquefaction process takes place by mixing the algae with a solution containing a low amount of acid (from 1 - 10 %) heated at temperatures of at least 80 ° C or higher for 30 minutes.[272]

Next, the process of saccharification is executed. This method breaks down starch compounds into the individual saccharide units of glucose. This process appropriately prepares the saccharide material in the form of glucose saccharides. Saccharification utilizes enzymes that get mixed into the broken down algae solution done at low temperatures. A very low amount of the enzyme gets utilized, correlating to less than 0.5 % of the total amount of starch contained in the pre-treated algal solution. This mixture then gets heated at around 50 – 60 °C for 30 minutes.[272] Next, the saccharides can remain with or be separated out from the rest of the broken down cell components before fermentation takes place. In other words fermentation can be executed without the separation of saccharides from the other biomass but it is usually preferable that some type of separation process happens before performing fermentation.

An experimental method involving the processes of liquefaction & saccharification of algae with subsequent fermentation by yeast cells

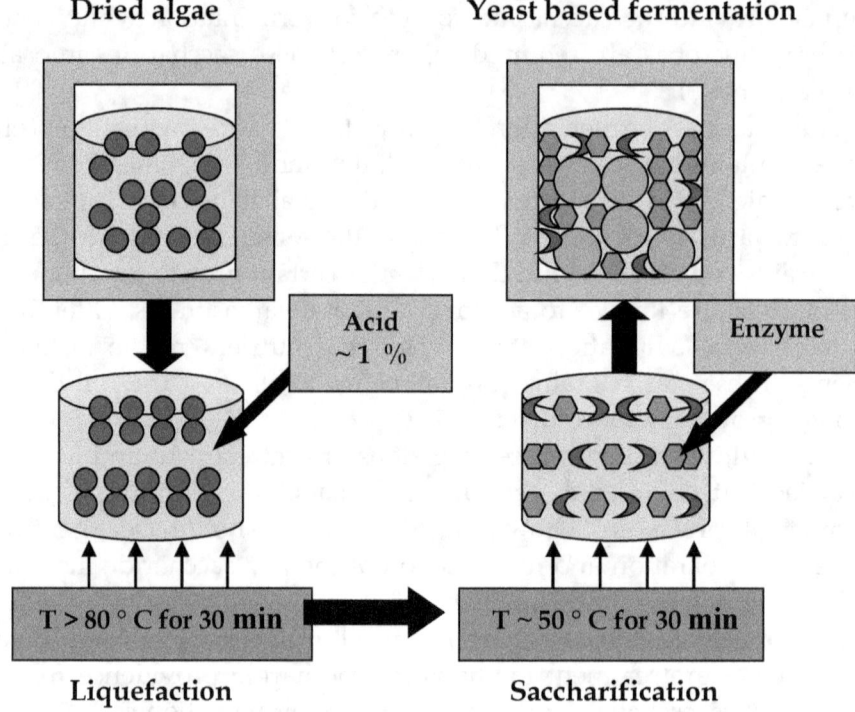

Figure 7-1 : The basic processing steps necessary towards breaking down and removing the saccharides from within algae are liquefaction and saccharification where the algae solutions are heated for 30 minutes with an acid and then an enzyme solution.[272]

The above mentioned pre-treatment processes are executed at low temperatures. Low temperature processing can be an advantage for biofuel manufacturing purposes in general. The advantage of low temperature processing is that it does not modify the biocomponents that were contained in the algae very much. This type of processing also tends to create a solution of biocrude as was discussed in chapter 3. Therefore, before fermentation with yeast cells takes place, the saccharides could be separated from the protein and lipid material. Low temperature processing methods have the potential to produce several biofuels at one time. For example biodiesel can be made from the lipid material extracted from the algae while the fermentation of saccharides from within them produces alcohol(s) like ethanol.

7.3 The biofuels of ethanol and renewable jet/diesel fuel can be made at one time with low temperature processing of cultivated algae

In the past, algae have been cultivated in photobioreactors located next to a coal plant for the purpose of producing several biofuels at one time. The idea of producing algae from the carbon dioxide emissions given off by power plants has been around for several decades but few if any companies have actually attempted to do it. One such project, headed by the company GreenFuel Technolgies, aimed to make the biofuels of ethanol and renewable diesel or jet fuel at a facility located in Arizona.[273] However, GreenFuel Technologies ceased algae production operations in 2009 due to financial difficulties.[274]

The company grew algae in plastic bags or thin plate photobioreactors housed within a greenhouse facility located very close to the coal plant. The photobioreactors directly took in carbon dioxide from flue gas emitted by the coal plant nearby. High density algae cultures were produced there since the flue gas contained carbon dioxide in high concentrations. The high concentration of algae produced resulted towards giving an excellent yield of biofuel made on the acre of land the algae was cultivated on. The company estimated that they could manufacture over 7,000 gallons of renewable hydrocarbon fuel and over 5,000 gallons of ethanol per year on an acre sized facility.[273]

The amount of algae required to produce 12,000 gallons of fuel is estimated to be over 1,000 tons of algae on an acre of land per year according to this model as shown in Figure 7-2. This figure also projects the respective amounts of lipids and carbohydrates utilized towards making the ethanol and renewable jet or diesel fuel. In addition, a large portion of the algal biomass (~ 45 %) is dedicated as a protein rich meal that can be utilized as animal feed. The amount of lipids, carbohydrates and proteins making up the 1100 tons or 100,000 kg of biomass correlate to a percentage composition of 25 %, 30 % and 45 % for each type respectively.

In order to make the biofuels of ethanol and renewable jet or diesel fuel, the saccharide, lipid and protein components must be separated from each other. For this reason, a low temperature treatment process is utilized before the separation of the biocomponents takes place. Low temperature processing is recommended due to its ability to preserve the chemical nature of the biocomponents contained in the harvested algae. Intact proteins are necessary to sell and market as animal food while unmodified saccharides are essential towards providing the correct glucose material for fermentation. Initial high temperature processing is avoided since nitrogen compounds in the proteins become modified into nitrogen based heterocyclic compounds while the saccharides get modified into compounds such as cyclic ketones, esters, furfurals and others.

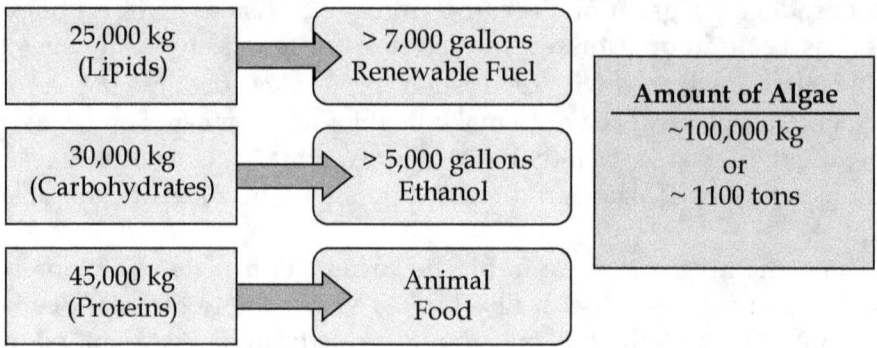

Figure 7-2 : Estimation by the author as to the amount of algae required for cultivation in order to make over 7,000 gallons of renewable fuel, 5,000 gallons of ethanol from yeast based fermentation and close to over 500 tons of animal feed on 1-acre of land based near a coal plant receiving carbon dioxide emissions from it.

After the low temperature treatment is executed, further processing separates the lipid, carbohydrates and proteins from one another. The specific processing methods associated with the separation of the biocomponents is more technically complicated and involved than can be discussed here. Regardless, after separation, the carbohydrate portion can be fermented in order to form ethanol while the lipid portion goes through hydrotreatment so that renewable jet or diesel fuel can be produced. It is also important to point out that the hydrotreatment manufacturing process can produce either renewable jet or diesel fuel.

7.4 Macroalgae is a very competitive energy crop that can produce several alcohols through indirect fermentation

Macroalgae, the large plant-like organisms that grow in ocean-side areas, include many varieties of kelp and seaweed. They are commonly encountered in various bays across the world where they grow in large colonies due to an excessive amount of nutrients that accumulate in such areas. These nutrients settle into the bays from agricultural and industrial chemical based runoff. The collection of macroalgae from bays and its resultant conversion into biofuels could be a valuable endeavor applied towards environmental land and sea based remediation.

Macroalgae can also be cultivated near seashore areas as an energy crop useful towards making biofuels and chemicals. They compete very well with other energy crops such as grasses. The purchase price of macroalgae is very reasonable at around $40 per ton.[275] In addition, its production cost per ton is comparatively around half the price of other crops like switchgrass. Macroalgae also has the advantage of producing a large amount of biomass on the land where it is cultivated. It tends to make about twice the amount of biomaterial per acre compared to other grass energy crops. For example, macroalgae produces around 24 tons of biomaterial per acre whereas switchgrass yields around 10 tons per acre per year.[276,277] These facts demonstrate that at least four times the amount of macroalgae can be made for the same price as it takes to cultivate and utilize grass type of energy crops.

Macroalgae as an energy crop is preferable to convert into biofuel for several other reasons. Primarily, it contains a large supply of saccharides that would be ideal towards producing alcohols through indirect fermentation. The saccharides can also be thermochemically converted into other hydrocarbons utilized for biofuel. Macroalgae differs markedly in biomass composition when compared to microalgae or other land based energy crops. It contains a different set of polysaccharides other than those encountered with regular plant or algae sources.

Macroalgae contains three types of polysaccharides or sugars that are potentially fermentable for microbes. These include laminarin (which is similar to cellulose), alginate and mannitol. Mannitol is actually a single saccharide alcohol unit whereas laminarin is made up of beta-glucan polymers that contain glucose units hooked together. Alginate consists of two uronic acids called mannuronic and glucoronic acids. The laminarin has the potential to be fermented by microbes but the other types of saccharides (alginate and mannitol) are more difficult to utilize for this purpose. Several options available for producing alcohols from macroalgae are covered below.

Butanol production from the majority of macroalgae saccharide sources : Past experiments have demonstrated the ability to make butanol from certain microbes by having them ferment some of the saccharide sources contained in the macroalgae. These saccharides include laminarin and mannitol. In these experiments pretreatment methods such as grinding and acid hydrolysis were required before fermentation was executed. These methods cause the laminarin and mannitol to be converted into the saccharide units of glucose and mannose. Using these methods, the DOE Pacific Northwest National

Laboratory (PNNL) prepared an aqueous extract consisting of the aforementioned saccharides that was later shown to produce butanol as well as organic acids upon fermentation with the bacterium *clostridium acetobutylicum*.[278]

Ethanol and isobutanol produced by genetically modified microbes : The best way to make ethanol from macroalgae is to utilize genetically engineered microbes that intake and metabolize all of the different polysaccharide sources contained in the macroalgae. Recently, research carried out by the company Bio-Architecture Laboratory (BAL) employs genetically modified *Escherichia coli* to allow all three saccharide sources obtained from macroalgae (*macrocystis pyrifera*) to be converted into ethanol.[277] In addition, the company has been developing genetically modified microbes that can produce the alcohol isobutanol from all of the saccharide sources in the macroalgae.[279] This type of alcohol fermentation builds upon past research work that demonstrated genetically modified *E. Coli* make appreciable amounts of isobutanol when they metabolize just glucose saccharides.[280] However, now these microbes can metabolize all types of saccharides utilized towards making the isobutanol. It is quite an accomplishment that such a company has the ability to produce several different types of alcohols with genetically modified microbes that are able to ferment a diverse array of saccharides contained in macroalgae.

This section presented the case that macroalgae are useful towards producing alcohols through indirect fermentation with genetically engineered microbes. However, there are other thermochemical based methods that can produce biofuels such as renewable diesel when applied to macroalgae. The next section discusses how certain thermal treatments that utilize easily accessible saccharides accomplish this feat.

7.5 The thermal processing of saccharides excreted from algae can produce a nice range of hydrocarbon compounds suitable for gasoline or diesel fuel

Saccharide sources obtained from algae do not have to go through fermentation in order to produce biofuel. These saccharides can produce hydrocarbons through hydrothermal treatment or a similar method known as aqueous phase processing. The saccharides contained in various biomass sources can be utilized to make a variety of hydrocarbons with aqueous phase processing. One possible source would include saccharides obtained from macroalgae. As mentioned in the last section, macroalgae contain three saccharide sources that include alginate, laminarin and mannitol. These saccharide sources are very ideal for producing hydrocarbons from hydrothermal (or aqueous phase) treatments. Therefore, it is preferrable to isolate the saccharides away from the other components of the macroalgae such as proteins and ash components (metals, etc) so that particular types of hydrocarbons can be produced.

Algae or microbes not only develop polysaccharides within their cells but they also excrete or form saccharides outside of their cells. Saccharides made by algae/microbes extracellularly are known as exopolysaccharides. Certain microbes also excrete amorphous cellulose from within their cells. Amorphous cellulose, different from crystalline cellulose found in lignocellulosic sources, would be easier to treat in order to break it down into individual glucose units. Thanks to recent developments it is now possible to make amorphous cellulose from algae cells. Researchers from the University of Texas have genetically modified cyanobacteria (algae) with the ability to make cellulose and saccharides such as glucose or sucrose.[281] In addition, these algae have the capability to assimilate carbon dioxide towards making the cellulose and saccharides.

Exopolysaccharides and excreted cellulose would be preferable to implement for aqueous phase processing since they do not require complicated separation methods to remove and isolate the saccharides from the algae prior to enforcing a hydrothermal treatment. However, the algae cells should be separated away from the saccharides contained in solution before this processing takes place. This is one of the biggest challenges to overcome concerning this production process. Regardless, the cultivation of algae for the purpose of producing amorphous cellulose or expolysaccharides would be ideal for hydrocarbon based biofuels utilizing aqueous phase (or hydrothermal) processing. Some methods for accomplishing this are described below.

Exopolysaccharide production from algae takes place under certain circumstances such as batch cultivation done for periods varying from two weeks to just over a month.[282,283] This period of cultivation is longer in duration than that which algae are accustomed to. During the time when the algae have gone past their optimum growth rate and have slowed down in division is normally when the exopolysaccharides are emitted from the algae.[283] The exopolysaccharides usually consist of a combination of pentoses and hexoses such as glucose, fucose, mannose, rhamnose, xylose, galactose and fructose.[283] These saccharides are a combination of five and six carbon sugars.

However, the distribution and amounts of saccharides vary with the algae species and length of the culture period. For example, spirulina contain large amounts of galactose, glucose and xylose while Phormidium spp. variety has considerable amounts of glucose, xylose, fucose and arabinose. In addition, more exopolysaccharides can be made by inputting a large amount of salt into the batch culture. It was demonstrated that a 4 %

exopolysaccharide concentration in the culture solution was produced by adding between 4-5 % salt in the media after 17 days with a certain species of cyanobacteria.[282] If the concentration in solution of exopolysaccharides reaches at least around 1 - 2 % from cultivation they may be further processed by a hydrothermal reactor that converts them into the chemical precursors required for the synthesis of fuel range hydrocarbons.[284,285]

The saccharides that undergo hydrothermal treatment can be converted into one of several different types of compounds. There are at least two types of approaches for converting the saccharides hydrothermally into hydrocarbon based biofuel compounds. One approach is to take the saccharides obtained from biomass and then put them through a medium to high hydrothermal treatment until they become organic acids.[285,286] The organic acids then get further converted into larger sized ketones through condensation reactions.[287] Afterwards another catalytic thermal process modifies the ketones into hydrocarbons. So in essence this method requires several aqueous phase processing stages in order to make the final hydrocarbon products from the original organic acids obtained from the saccharides.

The specific hydrocarbon compounds that can be synthesized from precursor chemicals derived from saccharides

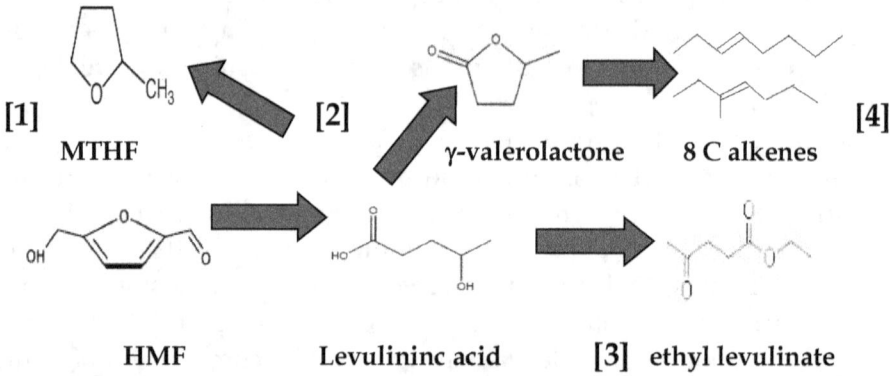

Figure 7-3 : These are some of the possible chemical synthetic pathways that are available towards the production of hydrocarbons usable for gasoline or diesel fuel obtained from levulinic acid. [289,291,292]

A second more common approach is to treat the saccharides with a medium temperature hydrothermal process. This type of process converts the saccharide of glucose into the chemicals of either hydroxymethylfurfural (5-HMF) or levulinic acid.[285,288] Oftentimes 5-HMF can turn into levulinic acid with further processing as is shown in Figure 7-3. Levulinic acid can then become modified through other synthesis methods in order to become biofuel type of hydrocarbons.

Figure 7-3 gives a more detailed depiction of how levulinic acid can be processed in order to make several types of biofuel compounds. For example, levulinic acid can be converted into either methyltetrahydrofuran (MTHF) [1], gamma valerolactone (GVL) [2] or esterified with alcohols to form fuel additive esters such as butyl or ethyl levulinate [3].[289,290,291] Methyltetrahydrofuran is a major component of the fuel known as P-Series Fuel, which is a vehicle fuel already approved by the Department of Energy.[291] Gamma valerolactone has the potential to go through further catalytic based synthesis to form a variety of eight carbon gasoline range double bonded branched alkenes [4] shown in Figure 7-3.[292] These branched alkenes are known to impart improved octane value to gasoline fuel when placed into it.

7.6 Potential biofuel compounds can be made through the natural excretion of organic compounds derived from algae

Under certain conditions, algae exist in natural lakes where a number of organic compounds are excreted from them. This situation is similar to the microbial mats introduced in chapter 5 where various growth conditions elicit the excretion of certain organic compounds. For example, a certain lake in the Middle East has been known to consist of a very high saline content as well as a lack of oxygen (i.e., anoxic conditions).[293] Algae that live in this lake have been known to excrete interesting organic compounds.

Experiments that isolate and grow these types of algae from the lake show the extent of such compound formations and excretions. One such experiment took algae from the lake and grew them on a microbial mat. The mat separated a number of algae into two growth layers (one on top and one on the bottom) as is shown in Figure 7-4.[294] The layers of algae excrete several types of organic compounds, two sets of which have implications towards renewable diesel fuel development. These compounds are n-alkane hydrocarbons of length C_{12} to C_{25} and fatty acids of carbon lengths C_{10} to C_{17}.[294]

These compounds normally do not exist in regular algae as separate singular entities but instead make up higher lipid structures such as triacylglycerides and phospholipids. Regular lipid metabolism tends to produce singular fatty acids that are then further converted into these complex lipid structures. However, in these circumstances, the fatty acids are made and excreted into solution instead of becoming further synthesized into triacylglycerides or phospholipids.

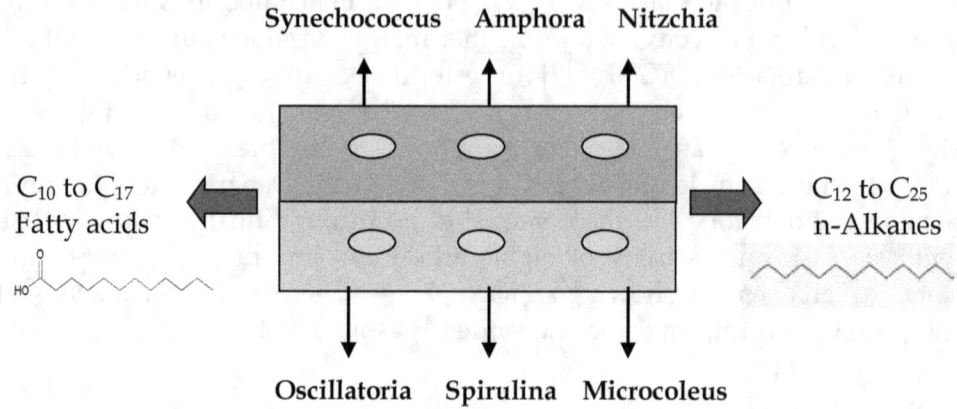

Figure 7-4: The cultivation of cyanobacteria isolated from a lake source that has high saline and low oxygen content for long periods of time can elicit the formation and excretion of a large number of compounds that include n-alkanes and fatty acids.[294]

Another natural system that produces free fatty acids and other related compounds that are excreted is the co-cultivation of *chlorella* with bacteria.[295] The type of compounds emitted are called **chlorellins**. These organic compounds are formed as a response to the algae growing together with bacteria. In other words, these compounds are released from the algae as a defense mechanism to bacteria that are being co-cultivated with them. A chemical analysis performed on chlorellin indicate that they contain a variety of hydrocarbon substances such as free fatty acids and other compounds that should be useful for biofuel production.[296] However, in order for them to be utilized for biofuel production they must be extracted from the culture solution. This is the general challenge regarding the cultivation of algae that emit organic compounds regardless of whether the algae are genetically modified or grow naturally.

Experimental methods of the past have assisted in the collection of such compounds. One such experiment utilized an adsorptive bag placed within the culturing vessel. It was able to gather the chlorellins together during the cultivation process as is shown to the left part of Figure 7-5.[297] In addition, another postulated method demonstrates that excreted compounds from algae (i.e., like chlorellins) could be gathered by solvent extraction involving a partitioned vessel. In the vessel, algae cells are grown in one half of it while the other part contains a hydrocarbon solvent as is shown in Figure 7-5.[298] The chlorellin compounds excreted find their way into the solvent portion of the vessel where they are taken up by it to be later separated from solution.

It is also known that other types of microbes and algae secrete similar types of hydrocarbon compounds when co-cultured with bacteria. These compounds are similar to chlorellin in that they contain free fatty acids along with polysaccharides and other smaller compounds. Since these hydrocarbons are a mixture of organic compounds they may be ideal as another biocrude material useful for further thermochemical processing methods such as hydrotreatment and liquefaction.

Possible extraction systems that should be able to separate chlorellins from the algae culture solution

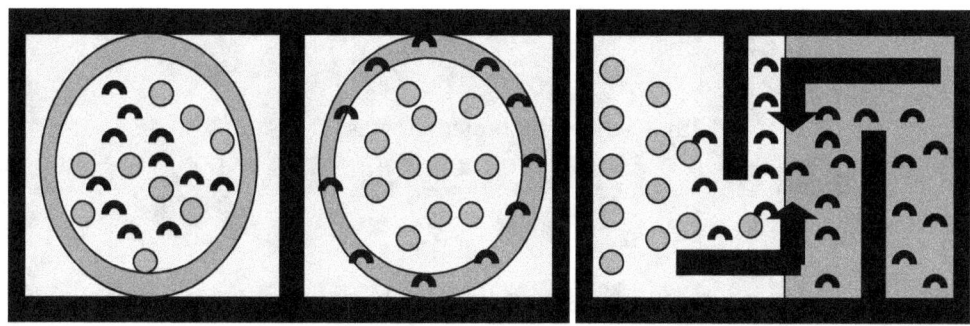

Figure 7-5 : Two possible extraction methods may be effective at isolating chlorellins from the culture solution using either solvent extraction contained on one half of the bioreactor or placing an adsorptive bag within a culture vessel that collects the chlorellins.[297,298]

7.7 Other types of excreted hydrocarbons useful for biofuel are made from genetically modified algae

The excretion of single fatty acids, alkanes and other related compounds can also take place with genetically modified algae or microbes. By adding specific enzyme systems to the microbes or algae, fatty acids get converted into singular alkanes, fatty alcohols, aldehydes or remain as free fatty acids.[307] This is shown in Figure 7-6. Alkanes are formed from an enzyme system that converts the fatty acids into linear hydrocarbons. Similar enzyme systems execute the formation of fatty alcohols or aldehydes from the fatty acids as well. The sizes (carbon number) of these compounds are correlated to the variance of fatty acids made within the cells themselves.

Fatty acids can range in size from 12 to 20 plus carbons in algae cells. The more common compounds found in microbes or algae are 16 or 18 carbon saturated and unsaturated fatty acids such as palmitic, stearic and oleic acids. The amount and distribution of fatty acids differs a bit between algae species in a manner similar to that of plant oilseeds. Algae, microbes and plant oilseeds contain a mixture of unsaturated (double bonded) and saturated (single bonded) fatty acids. Therefore, these types of compounds affect whether the fuel contains single or double bonded fuel compounds.

Fatty acid conversion into aldehydes, fatty alcohols, alkanes/alkenes or remaining as free fatty acids

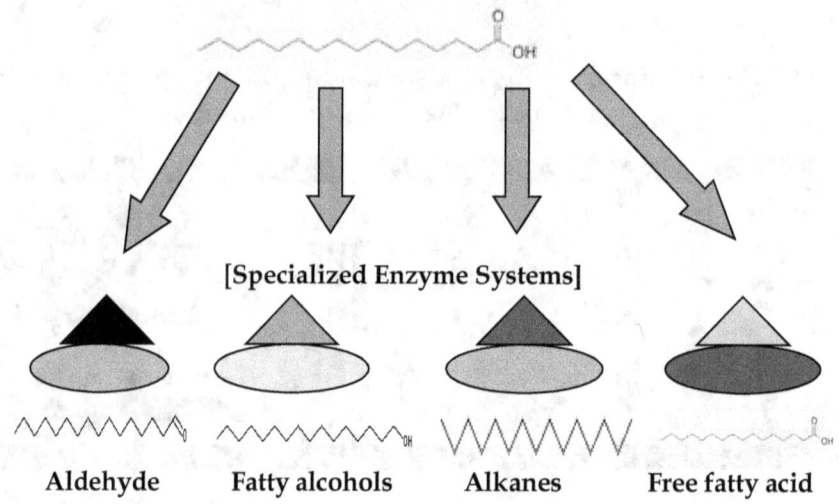

Figure 7-6: The inclusion of certain specialized enzyme systems in microbes or algae allow the production of a variety of hydrocarbons potentially useful as diesel fuel due to their size range.

Table 7-1 lists the type of conversion processes necessary to make certain biofuels from the production of alkanes, aldehydes, fatty alcohols and free fatty acids. As the table shows some compounds require conversion while others can remain unmodified after being extracted from the algae. Fatty acids of size lengths greater than 12 carbons are ideal for diesel fuel since this fits within their prescribed size range [ie ~12 to 24 carbons]. Of the four compounds described above, fatty alcohols and alkanes can be directly utilized as fuel compounds for renewable diesel fuel. Free fatty acids and aldehydes cannot be utilized as direct fuel compounds since they have some acidity value to them that makes them unsuitable for fuel use. However, free fatty acids can turn into biodiesel compounds through an esterification process. Both free fatty acids and aldehydes can also become renewable diesel compounds through reactions that take place with the hydrotreatment process. These reactions are decarboxylation and hydrogenation.

Another class of organic compounds that are emitted from genetically modified microbes is isoprenoids. These compounds are organized in groups of 5 carbons, so possible diesel fuel can be made from isoprenoids of 15 – 25 carbons in size.[300] The chemical structures of two common isoprenoids, isoprene and farnesene, are shown in Figure 7-7. Isoprene in the past has been successfully cultivated at high yields of 60 g/L and production rates of 2 g/L * hr by companies specializing in rubber manufacture.[300] The compound farnesene (a 15 carbon isoprenoid) has also been cultivated successfully with genetically modified S. Cerevisiae. The company Amyris in California has been manufacturing the compound from yeast cells both for vehicle fuel and extra value products.[301] Although isoprenoids are more commonly cultivated from yeast cells, in the future they may be cultivated from algae or bacteria.

Organic compounds emitted from microbes & algae that may require post modification treatment

Compound Type	Conversion Process
Free fatty acids (Too high acidity for direct use)	Esterification → [Biodiesel], Decarboxylation, hydrogenation → [Renewable Diesel]
Fatty alcohols (Direct use as fuel compounds)	No conversion process necessary [Renewable Diesel]
Alkanes/Alkenes (Direct use as fuel compounds)	No conversion process necessary [Renewable Diesel]
Aldehydes (Too high acidity for direct use)	Decarboxylation, Hydrogenation → [Renewable Diesel]

Table 7-1 : Fatty alcohols and alkanes/alkenes can be utilized without requiring modification. Aldehydes and free fatty acids due to their high acidity must be post processed using methods such as decarboxylation, acid catalysis or hydrogenation.

Some Isoprenoid compounds emitted by microbes

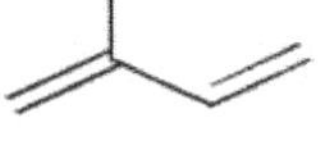

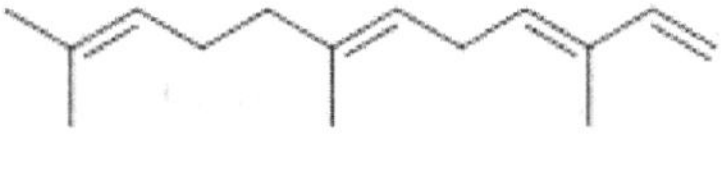

Figure 7-7 : Isoprenoids are compounds contained in 5 carbon increments such as isoprene and farenesene

The advantage of using algae or microbes to produce fatty alcohols, fatty acids, alkanes/alkenes, aldehydes or isoprenoids is that these compounds can be extracted out of solution. In addition, these types of cultures can be maintained so that a harvesting of the algae or microbes themselves is not required. This denotes that the algae or microbes can be cultivated for several generations to produce these organic compounds as opposed to using harvesting methods after a growth cycle just to obtain the fuel compounds of interest.

7.8 Direct ethanol production takes place with natural and genetically modified algae

This section covers material related to direct ethanol production from both natural and genetically modified algae. There are circumstances where natural algae produce appreciable quantities of ethanol. Certain types of cyanobacteria (algae) produce ethanol continually or require special cultivation conditions in order to excrete it at noticeable quantities. In the majority of cases, alterations to the normal cultivation of algae cause elevated levels of ethanol to be made. One such example includes the cultivation of the algae *Spirulina* utilized for making both ethanol and hydrogen. For other algae as well, specialized cultivation produces higher concentrations of both these compounds. Therefore, these conditions would be important to note since further biofuel applications could take place with the simultaneous production of ethanol and hydrogen. Hydrogen collected from the autofermentation of algae could be applied towards future manufacting, biofuel production and electrical power generation purposes as well.

This particular cultivation method with Spirulina requires a three stage process where the algae are grown and then get transferred consecutively to other bioreactors as shown in Figure 7-8. Much of this work has been performed by researchers from the University of Princeton.[302,303,304] The first stage grows algae with the lack of nutrients in the growth media and also implements high light intensity. The algae then get transferred to a second bioreactor that cultivates it in higher salt content and periodic changes in pH. Afterwards the algae get harvested and are placed into a vessel that grows them in dark anerobic conditions in order to induce an autofermentation process that makes the ethanol and hydrogen. The side products of acetate and formate are produced as well during autofermentation.

The combination of nutrient depletion, high light intensity, pH changes and high salt content result in a high accumulation of the polysaccharide glycogen that can be found in the spirulina cells at around 50 % total cell dry weight content. It is interesting to point out that high salt environments assist towards forming saccharide compounds. For example the compounds of trehalose and glucosyl-glycerol are saccharides found in algae cultivation done in high salt environments.[305]

The general cultivation process for inducing ethanol and hydrogen in Spirulina

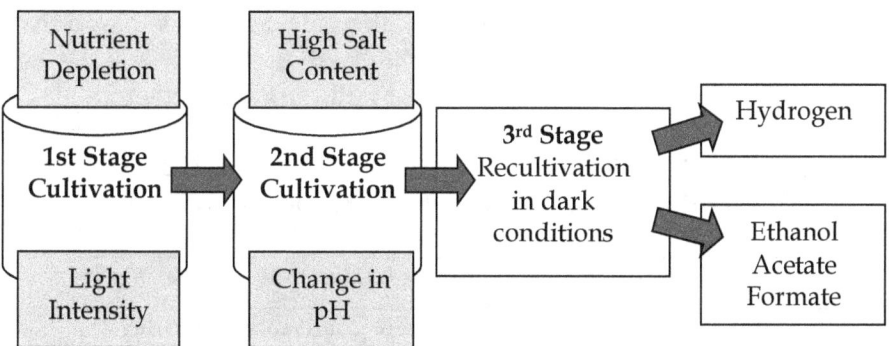

Figure 7-8 : Several growth cultivation stages and conditions are implemented in order to induce ethanol and hydrogen formation in the algae Spirulina.[302,304]

Another type of cyanobacteria that has the ability to produce ethanol throughout its cultivation cylcle without utilizing special growth conditions is the species *Cyanothece*. It is not necessary to cultivate these algae in separate bioreactors or isolate them in dark anaerobic conditions in order to induce ethanol formation. The natural genetic characteristics of the *Cyanothece* allow it to make ethanol in appreciable quantities under normal growth conditions. These characteristics are common to several different varieties of this species that have been collected in the past. Of particular importance is a *Cyanothece* variety that has been completely sequenced by researchers from Washington University.[306] Also, an additional 6 specific isolated strains of this cyanobacteria have been genetically sequenced by the DOE Joint Genome Initiative in order to give a very selective 'gene pool' set of this cyanobacteria.[307]

A list of the common genetic characteristics of *Cyanothece* that make them desirable for biofuel production are shown in Table 7-2. The synergistic effect derived from these characteristics allows for high amounts of both ethanol and hydrogen to be made. For example, a high nitrogen fixation rate is tied into the production of more carbohydrates during photosynthesis. More carbohydrates made during photosynthesis correlates to more ethanol produced later on from the algae. In addition, nitrogen fixation also helps to produce more hydrogen as a product.

The Cyanothece algae produce ethanol efficiently due to the effective operation of the internal enzymes they contain. This happens because intermediate compounds passed down through earlier metabolic processes are more directly converted into ethanol instead of being split into other byproducts. Therefore, the Cyanothece do not require additional enzymes necessary towards producing ethanol more exclusively. However, it does seem that the general trend for producing ethanol with algae takes place with growth systems that utilize genetically modified algae. This is due to the idea that genetically modified algae incorporate carbon dioxide as the carbon source instead of producing saccharides that build up within them.

Desirable metabolic based characteristics of *Cyanothece* strains useful towards efficient ethanol production

Have a high nitrogen fixation rate
High accumulation of carbohydrate grains during the day
Produces good yields of hydrogen during metabolism
Already contain the necessary enzymes for ethanol conversion that should not require genetic insertion for improved yield

Table 7-2 : A list of the known desirable metabolic characteristics contained in the several species of Cyanothece already fully genomically identified that could help make the process of ethanol (and hydrogen) production in algae an efficient process.[307,308]

Chapter 5 previously covered information regarding the Algenol Direct to Ethanol® process that makes ethanol from genetically modified algae assimilating carbon dioxide. The ability to metabolize carbon dioxide into ethanol takes place from the incorporation of specialized enzymes inserted into the algae. The enzymes inserted into genetically modified algae are *pyruvate decarboxylase* and *alcohol dehydrogenase*. Figure 7-9 demonstrates the difference between carbon dioxide and saccharide based metabolism resulting in ethanol production. The aforementioned metabolic processes involve the flow and conversion of intermediate compounds with or without genetically inserted enzymes.

The main metabolic intermediate responsible for making ethanol in all types of algae happens to be the compound of pyruvate. This compound tends to be the main branch point towards producing ethanol or other byproducts. Without specially added enzymes most of the compounds that get turned into pyruvate usually come from the metabolic process of glycolysis as is shown in part A of the figure. The normal enzymes then convert the pyruvate into a number of products that include acetate, formate and ethanol. However, algae species such as Cyanothece that mainly produce ethanol happen to be the exception to this general occurance.

A different result in producing mainly just ethanol from algae tends to be the case with algae that contain the specially added enzymes. With genetically modified algae, a large flow of compounds that eventually become pyruvate come from the two carbon dioxide assimilation cycles and the citric acid cycle as shown in part B of the figure. The formed pyruvate then becomes efficiently converted into ethanol through the specially added enzymes. This demonstrates that these added enzymes serve two functions. They cause the carbon flow to come from the citric acid cycle through carbon dioxide assimilation and also produce just ethanol instead of several chemical byproducts.

Metabolic comparison between algae that are genetically modified and those with normal enzymes

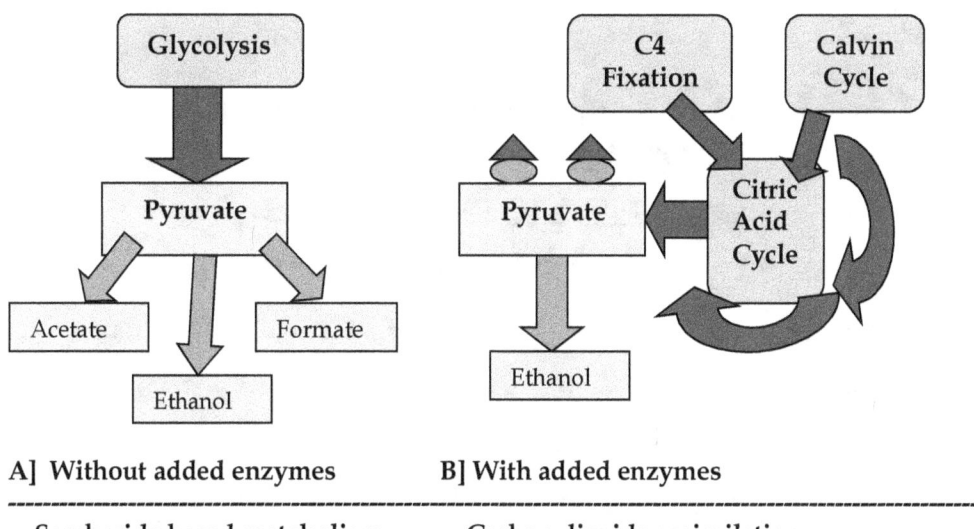

A] Without added enzymes   B] With added enzymes

Saccharide based metabolism   Carbon dioxide assimilation

Figure 7-9 : A) Natural algae with normal enzymes typically go through glycolysis based metabolism involving the breakdown of saccharides into ethanol and other byproduct formation. B) Genetically modified algae have enzymes inserted in them that cause carbon dioxide to be assimilated through the two different processes of C4 fixation and the Calvin Cycle. The intermediates then go through the citric acid cycle and become converted into pyruvate where the modified enzymes then directly convert pyruvate into ethanol.

7.9 Direct isobutanol production can also be done with genetically modified cyanobacteria (algae) cells

The alcohol isobutanol is a very attractive alternative as a vehicle fuel additive due to its high octane number and heating value. As a fuel additive it may be more conducive for vehicle use than ethanol due to its beneficial properties. These properties help to prevent detrimental wear to internal engine parts. For these reasons it would be prudent to pursue various methods of making isobutanol. Both direct and indirect methods for producing isobutanol from algae are presented in this chapter. In section 7.4 it was discussed that genetically modified microbes were able to metabolize saccharides in order to make isobutanol indirectly from macroalgae. A direct method for producing isobutanol from genetically modified algae is presented in this section.

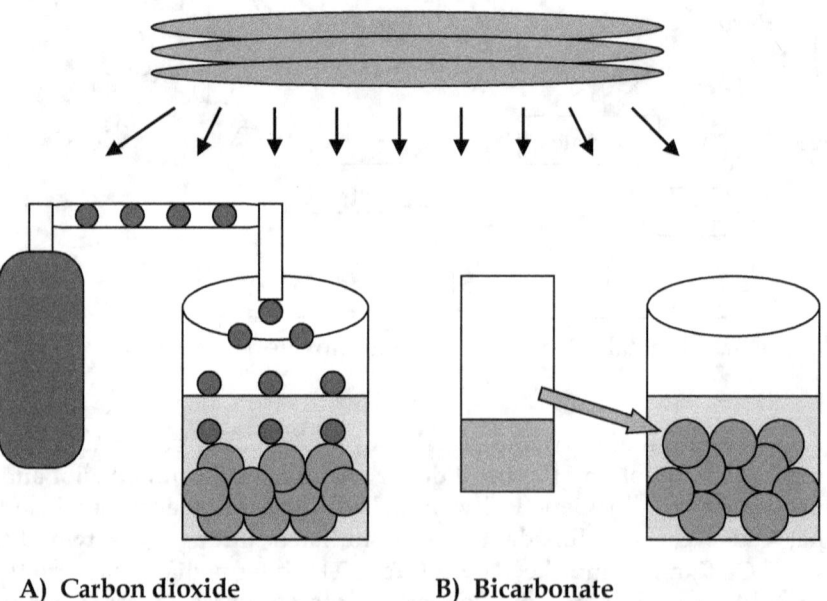

Figure 7-10 : A) Experimental cultivation of Synechococcus elongates was done in 1 Liter vessels with lighting and 5 % carbon dioxide bubbled into vessels with a total culture volume of 600 mL[317], B) The same lighting and culture volume was used with an alternate carbon source of bicarbonate in another experiment[309].

Researchers from the University of California have genetically modified a cyanobacteria algae called *Synechococcus elongates* to produce appreciable amounts of isobutyraldehyde.[309] Isobutyraldehyde can be further converted into isobutanol. This can either be done as a chemical reaction or performed biologically. Chemical conversion uses the process of reduction to do this. Biological methods involve the intake of isobutyraldehyde by enzymes of microbes that modify the compound into isobutanol.

The production of isobutyraldehyde performed by these researchers was done from the process of carbon dioxide fixation or alternatively metabolizing bicarbonate.[309] After genetically modifying the algae, they then grew it in 1 liter containers with 5 % carbon dioxide bubbled into the containers as is shown in Figure 7-10.[309] Alternatively, bicarbonate was added to the 1 liter vessels instead of carbon dioxide. These experiments were proven successful on a small scale in the laboratory but have yet to be executed in a larger scale setting.

Production rate comparison between ethanol or isobutyraldehyde made from algae

Compound type	Experimental or Estimated Production rate
Estimated ethanol from Direct to Ethanol® Algenol process	103 mg / L * day (Estimated Rate)**
Isobutyraldehyde from algae utilizing carbon dioxide	62 mg / L * day (Experimental Rate)
Isobutyraldehyde from algae Utilizing sodium bicarbonate	~ 140 mg / L * day (Experimental Rate)

Table 7-3 : The actual and estimated production rate from either ethanol production from the Algenol production model or isobutyraldehyde experimental production from algae.[309]

** Estimated production rate and yields per acre (8000 gallons per acre) based on the density of product, number of Algenol photobioreactors per acre, the volume of water per photobioreactor, ethanol yield per acre.

The large scale production of isobutyraldehyde may be possible in the future. The comparison between the Algenol Direct to Ethanol® method and the experimental isobutyraldehyde production demonstrates this point. The Algenol Direct to Ethanol® production model has already taken algae from the laboratory and made it suitable for production in an outdoor production model on a larger pilot plant scale. The estimated production rates of ethanol and isobutyraldehyde from these algae are shown in Table 7-3. Both production rates are similar in value. The results for laboratory based isobutyraldehyde production rate (62 mg / L * day) are somewhat in range to the estimated ethanol (from the Algenol process) production rate of 103 mg / L * day. In addition, note that if the algae are fed bicarbonate they produce isobutyraldehyde at close to double the rate (140 mg / L * day) compared to the carbon dioxide production scheme.

7.10 Multiple alcohols can be made simultaneously using genetically modified algae

It was shown earlier in this chapter that indirect methods involving the fermentation of algal saccharides oftentimes produces the alcohols of ethanol or butanol with natural microbes. Higher molecular weight alcohols can be made with algae utilizing direct or indirect production methods. However, their production usually requires the implementation of genetically modified microbes or algae as was just previously discussed concerning the production of isobutanol from algae. Currently, these types of higher molecular weight alcohols are made one at a time by a specific microbe or algae.

A major goal of modern day researchers is to produce a multiple set of alcohols simultaneously through genetically modified microbes or algae. Much of this work is still in the developmental stages or is theoretical in application currently. Regardless, the important message to convey to the reader in this section is to understand why a multiple set of alcohols can be produced at one time. Figure 7-11 outlines the various metabolic pathways that have the possibility of producing several types of alcohols at one time. These pathways include the production of alcohols such as ethanol, butanol, isopropanol, isobutanol, isopentanol and 2,3 butanediol.

All of the pathways originate from the chemical intermediate compound of pyruvate. Each branch emanating from pyruvate usually contains several other intermediates that are made until the alcohol of interest is produced. For example, isopropanol involves the formation of acetyl coA and then acetoacetyl coA that contribute to its production. Multiple alcohols produced from pyruvate in this fashion would require the insertion of the appropriate enzymes in the algae or microbes. Through this type of genetic modification, it may be reasonable to expect that a set of alcohols could be made simultaneously from one of the major branch points coming from pyruvate. Therefore, the production of alcohols such as 2,3 butanediol, isobutanol and isopentanol shown as the left side branch of Figure 7-11 would involve the inclusion of all the necessary enzymes.

Currently, these higher alcohols can be made through genetically modified microbes metabolizing simple saccharides. These experimental systems include the use of genetically modified *Escherichia Coli* that have been shown to produce the higher alcohols of butanol, isoproponal, isobutanol or isopentanol.[310] However, these microbes are designed to emit just one higher alcohol of interest at a time utilizing a set of inserted enzymes. These enzymes are functional towards directing the metabolic flow of intermediates towards the desired alcohol.

In addition, in these systems, the *E. Coli* usually only have the ability to incorporate the simple saccharide of glucose towards higher alcohol production. Therefore, it would be difficult to implement normal polysaccharide sources such as starch or cellulose for alcohol production. Both cellulose from lignocellulosic materials and starch obtained from algae require complicated pretreatment methods. Therefore, after pretreatment is executed the fermentation process itself is hindered due to the presence of interfering non-saccharide compounds that have not been completely removed from solution.

Two separate metabolic pathways with common branch points that could produce a set of alcohols

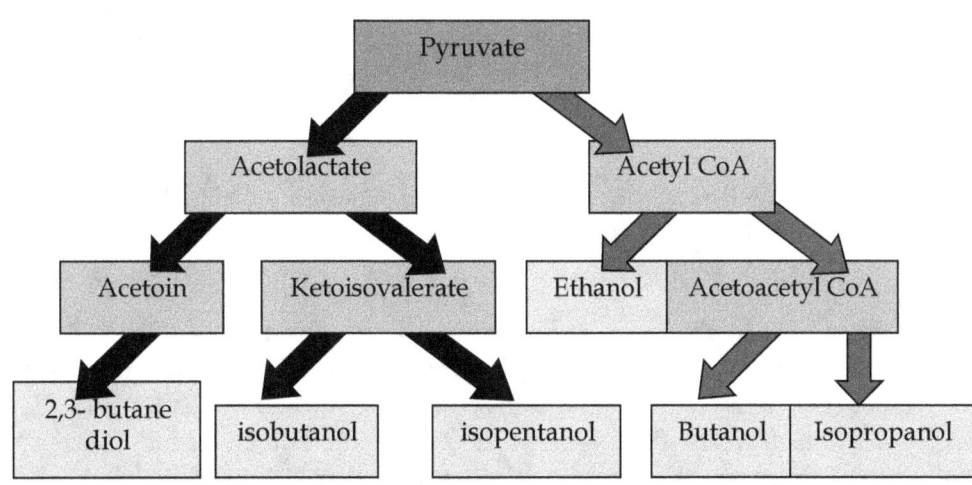

Figure 7-11 : Two separate metabolic branch points from pyruvate could potentially produce a set of multiple alcohols together.

However, developments in technology are now allowing for more pure forms of saccharides to be made from microbes or algae. In this way these microorganisms would provide a source of saccharides that could be utilized towards indirect production of alcohols from fermentation. One source of these saccharides happens to be amorphous cellulose normally excreted from microbes as was discussed in section 7.5. Amorphous cellulose, different from crystalline cellulose found in lignocellulosic sources, would be easier to break down into individual glucose units. In addition, this type of production from algae provides the simple saccharides of glucose and sucrose for further fermentation. In addition, these saccharides and cellulose are produced by algae that have been designed to assimilate carbon dioxide.

A theoretical two bioreactor model that produces alcohol from saccharides emitted by algae

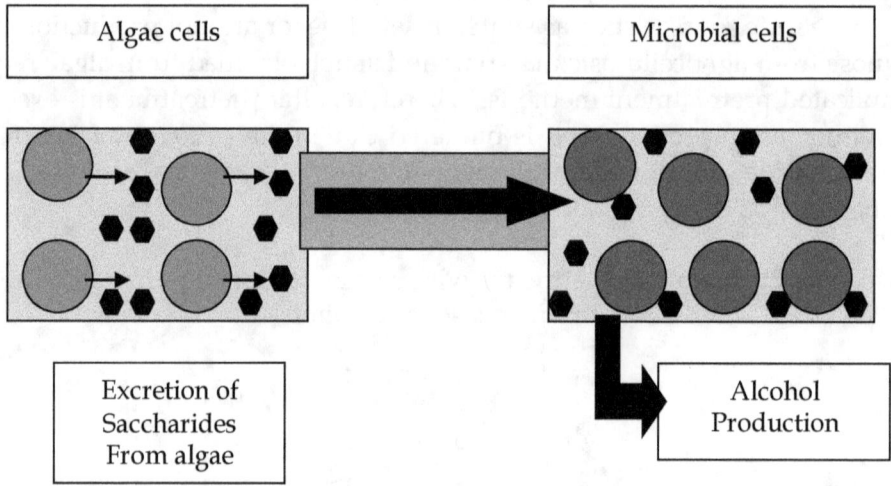

Figure 7-12 : A theoretical representation of a dual algae and microbial production system that makes alcohol(s).

A theoretical production system can be envisioned that totally incorporates carbon dioxide towards making both saccharides and alcohol(s) utilizing several bioreactors hooked together as is shown in Figure 7-12. One reactor would cultivate the algae with carbon dioxide in order to excrete saccharides while the other would ferment the saccharides for further alcohol production executed by either natural or genetically modified microbes. Regular microbes could produce the alcohols of either butanol or ethanol while genetically modified ones could make higher alcohols such as isopropanol, isobutanol, isopentanol and butanol.

Although this system may be suited towards producing alcohols with genetically modified microbes, it would be more ideal to directly make several alcohols at one time utilizing algae that assimilate carbon dioxide. Scientists believe that the direct, simultaneous production of several alcohols may be possible due to certain types of genetic modifications that could be done to algae. A theoretical model developed by scientists called the Photanol approach asserts that genetically modified algae can produce several alcohols at one time during the metabolic assimilation of carbon dioxide.[311] In addition, this model asserts that hydrogen can be made along with the alcohols as well.

Some of the specific cellular processes and chemical mechanisms directing the production of hydrogen and alcohols are depicted in Figure 7-13. However, due to their complexity, they are not described in much detail during this section. The interested reader can refer to the appropriate reference for more details.[311] In summary, the processes of oxidative photosynthesis and carbon dioxide assimilation help to form the intermediate compound of glyceraldehyde-3-phoshpate (GAP). This compound permits the fashioning of a number of different alcohol compounds when the correct enzyme systems are placed within the algae. GAP is high enough in the metabolic chain of processes that it can offer control over the possible types of products made later on.

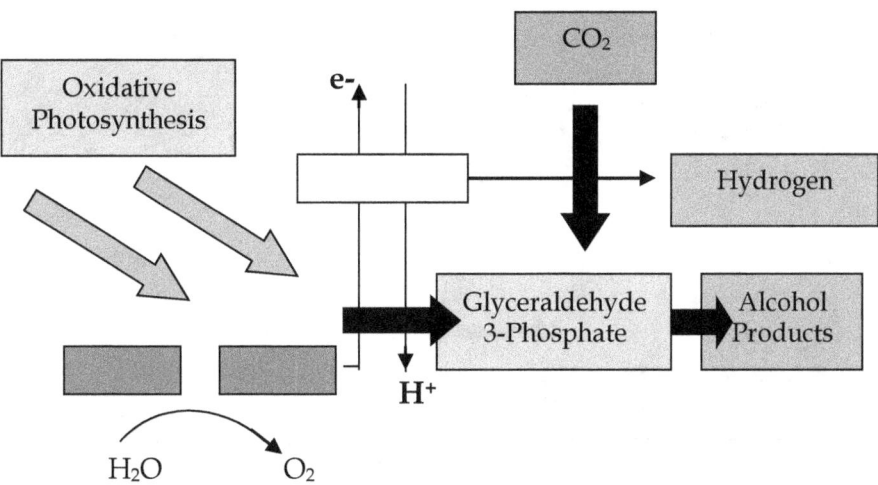

Figure 7-13 : The Photanol approach in theory combines oxidative photosynthesis with GAP based fermentation that produces alcohols, with the possible production of hydrogen.[311]

7.11 Algae can produce a high concentration of intracellular lipids for renewable diesel through nutrient alteration or heterotrophic fermentation

Lipids encountered in algae or microbes are normally contained within the cell membrane but are also found inside the cell as storage vesicles (lipid inclusions) as shown in Figure 7-14 below. Lipids known as triacylglycerides (TAGs) contained in the storage vesicles are more manageable in terms of processing for biofuel production. The structure of TAGs are shown to the right in Figure 7-14. A high concentration of these TAGs allows for an easier renewable diesel production process. Separation methods such as supercritical carbon dioxide extraction are applicable towards providing TAGs for eventual conversion into fuel. There are probably other methods not mentioned in this book that assist towards the separation and concentration of these TAGs for further biofuel production as well. The isolated TAGs can then be converted into renewable diesel through hydrotreatment or they can get turned into biodiesel through esterification.

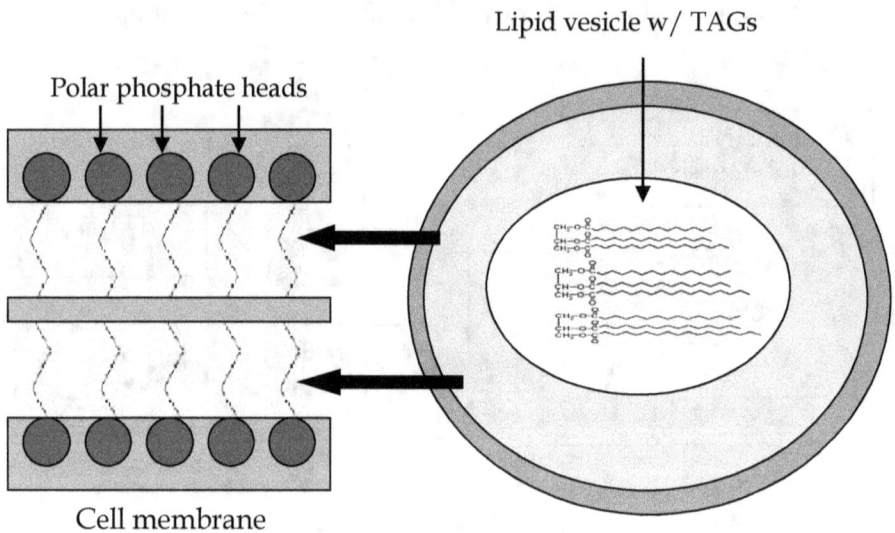

Neutral lipids (TAGs) contained in storage vesicles and polar phospholipids found within the cell membrane

Figure 7-14 : Lipids found within the cell in storage vesicles are neutral in chemical composition (non-polar) as opposed to the phospholipids encountered within the cell membrane. This non-polar nature makes them more manageable in terms of extraction and processing into biofuel such as biodiesel.

There are several ways in which algae can accumulate more intracellular lipids (TAGs) within them in order to be further processed into biofuel. When grown under normal conditions, algae contain up to 30 % lipid content. However, changes in growth conditions can boost the percentage of lipids contained in the algae during cultivation. Two approaches that have the potential to increase TAGs include **nutrient alteration** in cultivation media and **heterotrophic fermentation**. Nutrient alteration includes the methods of limitation, depletion or addition of chemicals to the media. However, nutrient removal usually results in lower growth rates and biomass production when done for further algae cultivation. Nevertheless, nutrient alteration appears to be a common way to achieve lipid accumulation since they can be applied towards many types of algae like the ones shown in Table 7-4.

Nutrient alteration increases the lipid percentage of most algae to around 50 % or higher. This applies for both freshwater and saltwater algae species. Table 7-4 shows the lipid percentage composition range for algae grown in normal or specialized conditions.[312] A large variation in lipid content applies for algae such as *chlorella protethecoides, botyrococcus braunii, nannochloris sp., dunaliella sp.,* etc. The lipid content of the majority of algae species undergoing nutrient alteration appears to accumulate to levels varying between 40 – 60 %. The main approach towards attaining higher lipid percentage within the algae cells has to do with altering the amount of nitrogen contained in the growth media. Nitrogen alteration of the growth media works since certain biomolecules rely on a steady supply of nitrogen in order to make up proteins while lipids mainly require a carbon source. Therefore compounds like lipids are made in preference to ones like proteins when there is a lack of nitrogen in the growth media.

To increase lipid content, several methods can be utilized towards altering the amount of nitrogen in the media fed to algae. One is sudden depletion of the nitrogen during regular cultivation. However, there are several disadvantages with this type of approach. An alternative preferred method is the gradual limitation of nitrogen that can be done in one of several different ways. One method includes nitrogen limitation by lowering the concentration of nitrogen in a second growth stage.[313] Another method involves the intermittent feeding (alternative cycles) of nitrogen during successive additions of nutrient media to the algae.[314] The best method may be nitrogen limitation through a series of dilutions given to the algae culture.[315] These methods may have the advantage of achieving normal growth rates and biomass concentration when compared to just straight nitrogen depletion.

Another way to increase the lipid content of algae rapidly is to have them undergo heterotrophic fermentation where a large amount of sugar or other carbon sources are fed to the algae within a short period of time. Heterotrophic fermentation denotes that the algae cultures are fed a carbon source while they are grown with little or no light. Oftentimes the algae are stirred, aerated, kept within a temperature range and fed more carbon compounds while undergoing this type of fermentation.[316] However, this method mainly applies to a limited set of algal species. This type of fermentation system oftentimes involves the cultivation of algae such as *chlorella protothecoides*.

Percentage composition range of lipids for certain freshwater and seawater algae

Dunaliella sp.	17 – 67 %	Chlorella protethecoides	14- 57 %
Nannochloris sp.	20 - 56 %	Neochloris oleoabundis	29 – 65 %
Botyrococcus Braunii	25 – 75 %	Phaeodactylum tricornutum	18 – 57 %
Schizochrytium sp.	50 – 77 %	Scenedesmus dimorphous	16 – 40 %

Table 7-4 : A list of several freshwater and seawater algae that have been known to accumulate lipids within their cells.[312]

Heterotrophic fermentation itself is adaptable to a number of different carbon sources that include acetate, glucose (i.e., saccharides) and glycerol.[317,318] Glucose based fermentation is the most common one encountered. Therefore, during fermentation the carbon source (i.e., glucose) is fed regularly to the algae until they reach a high cell density (> 15 g/L) and have a large lipid content (> 50 %).[317] Some fermentation experiments have produced yields of up to 100 g/L of algae cells when high amounts of glucose are initially and periodically added.[316] A higher cell density in combination with a higher intracellular lipid content makes it that much more easier to harvest and process the algae into biofuel.

Heterotrophic fermentation appears to be the type of technology utilized by one of the major companies making renewable diesel with algae. The company Solazyme has been quite successful in producing biofuel from algae using heterotrophic fermentation that they have already qualified for integrated biorefinery support from the government. In summary, heterotrophic fermentation appears to be one of the best ways to achieve a high triacylglyceride accumulation within the algae cells in order to better assist the various processing methods necessary to make biofuel.

REFERENCES:

271. Microalgae – novel highly efficient starch producers – Biotechnology and Bioengineering Vol 108 No 4 pgs 766 – 776 [2011] by I. Branyikova, B. Marsalkova, J. Doucha, T. Branyik et al

272. **Enzymatic pretreatment of chlamydomonas reinhardtii biomass for ethanol production** – Bioresource Technology Vol 101 pgs 5330 – 5336 [2010] by SP Choi, MT Nguyen, SJ Sim

273. **Chapter 8 Who is producing algae? In Green Algae Strategy – End oil imports and engineer sustainable food and fuel** – by M. Edwards [2008]

274. **Greenfuel Technologies closing down** – by M. Kanellos [May 2009] – http://www.greentechmedia.com/ -- accessed May 2014

275. **Macroalgae Butanol – Categorical Exclusion Determination Form** – US DOE ARPA-E Program [2009]

276. **Breaking the biological barriers for cellulosic ethanol : A joint research agenda** – US DOE EERE [2006] by J. Houghton, S. Weatherwax, J. Ferrell

277. **An engineered microbial platform for direct biofuel production from brown algae** – Science Vol 335 No 6066 pgs 308 – 313 [2012] by AJ Wargacki, E. Leonard, MW Win et al

278. **Acetone butanol fermentation of marine macroalgae** – Bioresource Technology Vol 108 pgs 305 – 309 [20xx] by MH Huesemann, LJ Kuo, L Urquhart, GA Gill, G Roesijadi

279. **Unlocking seaweed's next-gen crude : sugar** – The New York Times Environment Green Blog [Jan 2012] by J. Garthwaite – http://green.blogs.nytimes.com/2012/01/23/unlocking-seaweeds-next-gen-crude-sugar/ - accessed May 2014

280. **High flux isobutanol production using engineered Escherichia Coli : A bioreactor study with in situ product removal** – Applied Microbiology and Biotechnology Vol 86 pgs 1155 – 1164 [2010] by MR Connor, AF Cann, JC Liao

281. **New source for biofuels discovered at the University of Texas at Austin** – [Apr 2008] – http://www.utexas.edu/news/2008/04/23/biofuel_microbe/ -- accessed May 2014

282. **Physiological characterization and stress induced metabolic responses of Dunaliella salina isolated from salt pan** – Journal of Industrial Microbiology and Biotechnology Vol 35 No 10 pgs 1093 – 1101 [2008] by A. Mishra, A. Mandoli, B. Jha

283. **Extracellular polysaccharides from Ankistrodesmus densus (Chlorophyceae)** – Journal of Phycology Vol 34 pgs 638 – 641 [1998] by BS Paulsen, T Aslaksen, CS Freire-Nodi et al

284. **Chemical composition and production of exopolysaccharides from representation members of heterocystous and non-heterocystous cyanobacteria** – Phytochemistry Vol 52 pgs 639 – 647 [1999] by B. Nicholaus, A. Panico, L. Lama, I. Romano

292. **Dissolution and hydrolysis of cellulose in subcritical and supercritical water** – Industrial and Engineering Chemistry Research Vol 39 pgs 2883 – 2890 [2000] by M. Sasaki, Z Fang, Y Fukushima et al

285. **Hydrochloric acid catalyzed levulinic acid formation from cellulose : data and kinetic model to maximize yields** – AlCHE Journal Vol 58 pgs 236 – 246 [2012] by J. Shen, CE Wyman

286. **Hydrothermal upgrading of biomass to biofuel : Studies on some monosaccharide model compounds** – Carbohydrate Research Vol 339 pgs 1717 – 1726 [2004] by Z. Srokol, AG Bouche, A Van Esterik, RCJ Strick et al

287. **Integration of C-C coupling reactions of biomass derived oxygenates to fuel grade compounds** – Applied Catalysis B: Environmental – Vol 94 pgs 134 – 141 [2010] by EL Gurbuz, E. Kunkes, J. Dumesic

288. **Catalytic conversion of cellulose to chemicals in ionic liquids** – Carbohydrate Research Vol 346 pgs 58 – 63 [2011] by F. Tao, H. Song, L. Chou

289. **Solid acid catalyzed glucose conversion to ethyl levulinate** – Applied Catalysis A: General Vol 397 No 1-2 pgs 259 – 265 [2011] by L. Peng, L. Lin, J. Zhang, J. Shi, S. Liu

290. **A sustainable process for the production of gamma-valerolactone by hydrogenation of biomass derived levulinic acid** – Green Chemistry Vol 14 pgs 688 – 694 [2012] by AMR Galleti, C. Antonetti, V. De Luise, M. Martenelli

291. **Production of levulinic acid and use as a platform chemical for derived products** – Resources, Conservation and Recycling Vol 28 pgs 227 – 239 [2000] by JJ Bozell, L Moens, DC Elliot, Y Wang et al

292. **Integrated catalytic conversion of gamma-valerolactone to liquid alkenes for transportation fuels** – Science Vol 327 pgs 1110 – 1114 [2010] by JQ Bond, DM Alonso, D Wang, RM West, JA Dumesic

293. **Solar lake (Sanai) distribution of photosynthetic microorganisms and primary production** – Limnology and Oceanography Vol 22 pgs 609 – 620 [1977] by Y Cohen, WE Krumbein, M Shilo

294. **Excretion products of algae and their occurance in Solar Lake. Taba, Egypt** – International Journal of Environmental Health Vol 9 No 3 pgs 233 – 243 [1999] by MI Badawy, HF Abou-Waly, GH Ali

295. **Chlorellin, an antibacterial substance from chlorella** – Science Vol 99 No 2574 pgs 351 – 352 [1944] by R. Pratt, TC Daniels, JJ Eiler et al

296. **Qualitative analysis secretions with multiple mass spectrometric platforms** – Journal of Chromatography A - Vol 1244 pgs 139 – 147 [2012] by T. Kind, JK Meisson, D. Yang, F. Nocito et al

297. **Studies on chlorella vulgaris XI – Influence on growth of chlorella of continuous removal of chlorellin from the culture solution** – American Journal of Botany Vol 31 pgs 418 – 421 [1944] by R. Pratt

298. **Algal physiology; a catch all [Presentation]** – by R. Sayre – http://www.nrel.gov/biomass/pdfs/sayre.pdf - accessed May 2014

299. **Microbial production of fatty acid derived fuels and chemicals from plant biomass** – Nature Vol 463 pgs 559 – 562 [2010] by EJ Steen, Y. Kang, G. Bokinsky, Z. Hu, A. Schirmer, A. McClure et al

300. **Microbial production of isoprenoids** – Process Biochemistry Vol 46 pgs 1703 – 1710 [2011] by SS Chandran, JT Kealey, CD Reeves

301. **Amyris | Products** – http://www.amyris.com/Products/172/Products -- accessed May 2014 (no longer available)

302. **Contribution of a sodium ion gradient to energy conversation during fermentation in the cyanobacterium Athrospira (spirulina) maxima CS-328** – Applied and Environmental Microbiology Vol 77 pgs 7185 – 7194 [2011] by D. Carrieri, G. Ananyev, O. Lenz et al

303. **Optimization of metabolic capacity and flux through environmental cues to maximize hydrogen production by the cyanobacterium Arthrospira (spirulina) maxima** – Applied and Environmental Microbiology Vol 79 No 19 pgs 6012 – 6113 [2008] by G. Ananyev, D. Carrieri, GC Dismukes

304. **Boosting autofermentation rates and product yields with sodium stress cycling : Application to production of renewable fuels by cyanobacteria** – Applied and Environmental Microbiology Vol 76 No 19 pgs 6455 – 6462 [2010] by D. Carrieri, D. Momot, IA Brasg, G Ananyev et al

305. **Stabilization of model membranes during drying by compatible solutes involved in the stress tolerance of plants and microorganisms** – Biochemical Journal Vol 383 pgs 277 – 283 [2004] by DK Hincha, M. Hagemann

306. **Department of Energy funds cyanobacteria sequencing project – making green mats of ethanol** – Washington University in St Louis News and Information [Oct 2006] by T. Fitzpatrick

307. **Better living through Cyanothece – unicellular diazotrophic cyanobacteria with highly versatile metabolic systems** [2010] by LA Sherman, H. Min, J. Toepel, H. Pakrasi

308. **Novel metabolic attributes of the genus Cyanothece, compromising a group of unicellular nitrogen fixing bacteria** – American Society for Microbiology Vol 2 No 5 e00214-11 [2011] by A. Bandyopadhyay. T.Elvitogala, E. Welsh, J. Stockel et al

309. **Direct photosynthetic recycling of carbon dioxide to isobutyraldehyde** – Nature Biotechnology Vol 27 No 12 pgs 1177 – 1180 [2009] by S. Atsumi, W. Higashide. JC Liao

310. **Non fermentative pathways for synthesis of branched chained higher alcohols as fuels** – Nature Vol 451 pgs 86 – 89 [2008] by S. Atsumi, T. Hanai, JC Liao

311. **Alternative routes to biofuels : Light-driven biofuel formation from CO_2 and water based on the photanol approach** – Journal of Biotechnology Vol 142 pgs 87 – 90 [2009] by KJ Hellingwerf, MJ Teixeira de Mattos

312. **Microalgae : A promising feedstock for biodiesel** – African Journal of Microbiology Research Vol 3 No 13 pgs 1008 – 1014 [2009] by X. Deng, Y. Li, X. Fei

313. **Chemical profiles of microalgae with emphasis on lipids** – Final Report submitted to SERI by JR Benneman, DM Tillett, Y. Suan et al – accessed on internet

314. **Influence of nitrate feeding on carbon dioxide fixation by microalgae** – Journal of Environmental Science and Health : Part A – Vol 41 pgs 2813 – 2824 [2006] by HF Jin, BR Lim, K Lee

315. **Modelling neutral lipid production by the microalgae Isochrysis aff. Galbana under nitrogen limitation** – Bioresource Technology Vol 102 pgs 142 – 149 [2011] by F. Mairet, D. Bernard, P. Masei et al

316. **Production of high density chlorella culture grown in fermenters** – Journal of Applied Phycology Vol 24 No 1 pgs 35 – 43 [2012] by J. Doucha, K. Livansky

317. **Oil accumulation via heterotrophic/mixotrophic chlorella protothecoides** – Applied Biochemistry and Biotechnology Vol 162 No 7 pgs 1978 – 1995 [2010] by T. Heredia-Arroyo, W. Wei, B. Hu

318. **Heterotrophic culture of chlorella protothecoides in various nitrogen sources for lipid production** – Applied Biochemistry and Biotechnology Vol 160 pgs 1674 – 1684 [2010] by Y Shen, W Yuan, Z Pei, E Mao

GLOSSARY:

Acetic acid – A two carbon carboxylic acid. It is the most common organic acid (carboxylic acid) produced during fermentation. This organic acid then gets converted into the alcohol ethanol through other thermochemical unit processes executed within a biofuel refinery. It is the main organic acid produced during carboxylate counter-current fermentation and fermentative acetic acid conversion.

Acetogens – Specialized types of microbes that are able to produce acetic acid through anaerobic respiration. They can utilize both saccharide and carbon dioxide to produce the ethanol from fermentation. These are the types of microbes that are implemented for fermentative acetic acid conversion by the company Zeachem.

Acid Gas Removal (AGR) system – A piece of equipment utilized for the removal of certain impurity gases found in the synthesis gas product stream. They are removed so that they do not damage the metal catalysts contained in the synthesis stage. Oftentimes they remove gases such as hydrogen sulfide or carbonyl sulfide.

Acid Value (biodiesel) – A value that measures the amounts of free fatty acids contained in a biodiesel sample being tested to pass certification standards.

Advanced Biofuels – A biofuel category classified by the Renewable Fuel Standards that define it as a biofuel made from renewables other than corn starch. This mainly includes certain cellulosic biofuels such as renewable diesel. It is expected that 21 billion total gallons be made from advanced biofuels by 2022.

Air to fuel ratio – The mass or stoichiometric ratio of air to fuel introduced into an engine cylinder before combustion occurs. The efficient operation of an engine requires an optimum air to fuel ratio value. This is due to the fact that efficient combustion needs a certain amount of oxygen that is volatized with the fuel mixture.

Air Separation Unit (ASU) – A piece of equipment that can separate oxygen from the air towards use in a pressure blown oxygen gasifier.

Alcohol dehydrogenase – One of the several enzymes found in yeast cells that allows for the conversion of pyruvate into ethanol. One of the other enzymes, *pyruvate decarboxylase,* converts pyruvate into acetaldehyde while *alcohol dehydrogenase* helps to convert the acetaldehyde into ethanol.

Aldehydes – An organic compound class that is made up of a carbonyl group with a hydrogen attached at one end of it (H-C=O). Aldehydes are compounds found in bio-oil and potentially can be produced from fermentation. Once isolated they are not suitable for direct use as a fuel compound but can be chemically modified somewhat into other types of compounds that are.

Algae based ethanol production (via assimilation of carbon dioxide) – Genetically modified algae can produce ethanol through the assimilation of carbon dioxide. The carbon dioxide goes through several metabolic cycles and then gets converted into ethanol with the assistance of two inserted enzymes (*pyruvate decarboxylase* and *alcohol dehydrogenase*). Algae based ethanol is one of the alternative ethanol production methods being utilized by integrated biorefineries at a pilot plant scale level. High levels of ethanol can also be produced by natural algae such as *Spirulina* or *Cyanothece* using normal or specialized cultivation conditions.

Alkali minerals – Earth based metal compounds contained originally in the lignocellulosic or other biomass that are separated from it during thermochemical gasification. They become part of the char material or are dispersed with the synthesis gas. They are also the minerals components of metal based catalysts utilized in thermal gasification.

Alkanes – Organic compounds that can found in a number of configurations, structures and sizes. They include straight chained, cyclic and branched compounds. However, they do not contain double bonds in them. The term paraffin is very similar in application.

Alkenes – Organic compounds that contain double bonds in them. They oftentimes include both branched and straight chained compounds. The term olefins have the same meaning.

Alkyl esters – The ester type of compounds that make up biodiesel. These esters are formed upon the combination of a fatty acid that reacts with an alcohol requiring the assistance of a hydroxide type of catalyst. Methyl esters are a type of alkyl ester. Usually just one kind of alkyl ester is produced depending upon the type of alcohol utilized. For example, ethyl esters are made from ethanol. However, a range of different alkyl ester compounds are produced since there are different types of fatty acids contained within the cells of microorganisms, oilseeds or animal fats.

Alkylation – Reactions that take place during fuel refining where small hydrocarbon groups from certain compounds get transferred to other compounds. This usually results in the formation of fuel compounds that consist of branched alkane or alkene types of hydrocarbons.

Alternative ethanol production – The manufacture of ethanol performed in large scale refineries that do not usually implement conventional fermentation methods. Conventional fermentation normally includes both lignocellulosic and corn starch based fermentation. There are many different types of alternative production methods available towards making ethanol at least on the pilot plant level. These include synthesis gas fermentation, thermal gasification, fermentative acetic acid conversion and algae based ethanol production from carbon dioxide assimilation .

American Society for Testing and Materials (ASTM) – An organization that produces and publishes standards for acceptable manufacturing or product testing applications and practices. Regarding biofuels manufacture, some specific standards have been adopted by the government. Some examples in this book include the ensurance of quality biodiesel manufacturing (ASTM D975 & D7647) along with the testing of a fuel's octane number (ASTM D2700).

Amino acids – The basic compound units that make up proteins. They are compounds made up of amino, carboxyl and other side chain functional groups. There are over 20 different individual amino acids that can be connected together in different ways to form complex protein structures.

Amorphous cellulose – Cellulose contained in a gelatinous, unorganized form that can result in easier degradation with enzymes due to its accessibility and nature. This type of cellulose is often produced by bacteria. It has also been produced by algae that have been genetically modified.

Anaerobic Digestor (AD) – An energy facility that produces electrical energy from the generation of biogas or its further conversion into hydrogen. The biogas is produced from microbes that anaerobically digest biomass material that usually consists of renewable wastes oftentimes from crop residues, animal wastes or food processing wastes.

Anaerobic mesophiles – Microbes that go through anerobic digestion and grow best under normal temperature conditions (not too cold or hot). These types of microbes are utilized for synthesis gas fermentation.

Anhydrous ethanol – Ethanol that contains little to no water in it. This means that it usually has greater than 99 % ethanol when it is sold as a product. This is the type of ethanol utilized as a fuel additive in gasoline. The production of anhydrous ethanol requires additional processing after distillation such as dehydration therefore, it is normally more expensive to purchase than the hydrous version.

Anoxic – An environment that contains little to no amounts of dissolved oxygen in it. Such conditions can be toxic to certain organisms (i.e., microbes) that grow in aqueous environments.

Aqueous phase processing (reforming) – A hydrothermal processing method that effectively converts saccharide material into useful chemicals and/or fuel compounds. It accomplishes this through the efficient design of reactors that contain specialized catalytic materials carried out at specific temperature and pressure conditions. When the process also makes hydrogen gas that can be later utilized to help further refine the made chemicals it is called **aqueous phase reforming**.

Aromatic hydrocarbons – Compounds that contain benzene molecules hooked together or with attached side groups. The monolignol compounds in lignins are also considered as aromatic hydrocarbons

Aqueous fraction of bio-oil – The compounds found in bio-oil can be separated into water and hydrocarbon soluble portions. The water soluble or aqueous portion contains many organic compounds that are polar such as polysaccharides, organic acids, aldehydes and alcohols.

Autofermentation – The manipulation of growth cultivation conditions that allow for the production of certain metabolic products from the breakdown of stored carbon based compounds within the microbe or algae. These cultivation conditions usually take place with little to no light and in the absence of air – otherwise known as **dark anaerobic conditions**.

Batch cultivation or processing – A production or cultivation process that takes place in a confined reactor for a specified period of time. It is different from a semi-continuous or continuous method where products are being made and removed throughout the duration of processing. Instead biomass or products are made after a given period of time and then removed by some means. Afterwards another 'batch' production cycle takes place with the introduction of fresh materials into the reactor.

Billion ton study – A joint US government study conducted in 2004 by the USDA and DOE that attempted to classify and enumerate renewable feedstock sources that are available on a yearly basis that can be applied towards conversion into fuel, energy or products. The study was valuable in the fact that it attempted to define and establish practices that can be executed by farmers, businesses or government entities that would allow high yields of these renewables to be available for future use.

Biocrude – The end products from the thermochemical processing of biomass usually taking place in aqueous environments. Biocrude could consist of either bio-oil produced from liquefaction or broken down biocomponents that are made from a low temperature hydrothermal processing method. These biocomponents are usually lipids, polypeptides (proteins) and saccharides. Biocrude similar to petroleum based crude should have the potential to be formed into a number of different vehicle fuels and chemical byproducts.

Biomass based diesel – One of the biofuel categories classified by the Renewable Fuel Standards that defines diesel produced from biomass sources that do not include lignocellulosics. This does include diesel made from oilseeds or from algae. Therefore, this includes biodiesel. There is a cap on biomass based diesel production set at 1 billion gallons per year that remains until 2022 according to RFS2.

Biomass Conversion Facility (BCM) – Facilities strategically located near farms or other areas that produce sufficient quantities of biomass material. The goal given by the US government is to locate these facilities within 50 – 100 miles of most major farms or places that contain a major feedstock source (i.e., like forestry waste). A BCM can apply both towards a refinery that produces a fuel/product or a processing center that fashions the biomaterial into a form to where it can be stored or transported easier to a refinery.

Biomass Crop Assistance Program (BCAP) – A financial assistance program available from the Farm Service Agency offered to agricultural and private forestry land owners that plan on cultivating and harvesting energy crops that can be later delivered to biomass conversion facilities. The program allows for matching payments made to the landowners for crops delivered to BCM's successfully.

Bio-oil – A viscous liquid containing hundreds of different organic compounds that mainly consist of oxygenated organic compounds. The bio-oil is produced from pyrolysis or liquefaction. The product cannot be directly utilized as a vehicle fuel since it has many adverse qualities and properties. It must go through an upgrading process such as hydrotreatment in order to be suitable for vehicle fuel use.

Black liquor – The liquid product obtained after the digestion process that takes place during the kraft mill wood pulp manufacturing process. It contains lignins, hemicellulose and other chemicals. It is used as a source for producing electrical and heat energy for the pulp mill. It can also be utilized towards the production of biofuel.

Break even price (for ethanol biorefinery) – The relationship between the price of corn and the commodity price of ethanol. This theoretically sets a break even point where profits are no longer earned from the production of ethanol from corn if the price of corn rises higher than the break even price.

Brown (waste) grease – A renewable waste obtained from food preparation or processing that normally happens in places such as restaurants. Unlike waste cooking oil, brown waste grease is obtained from grease traps located near sinks. Similar to waste cooking oil this renewable waste can be converted into biodiesel and may yield just as large a supply as waste cooking oil does.

Bushel – A term used to describe the amount of a crop sold, traded, etc. A bushel varies by weight somewhat dependent upon the crop but is normally from 40 – 70 pounds per unit of bushel.

BTX (Benzene, Toluene, Xylene) – The three most common aromatic compounds that are utilized for the further manufacture of products. Petrochemical based manufacturing oftentimes initially produces these compounds together as a set. Therefore, they can be further processed together in order to become refined chemical ingredients used to make consumer or industrial products. It is also interesting to note that biomass types of production methods are producing these types of aromatics together as well.

Calcium carbonate – A chemical utilized in the carboxylate counter-current fermentation method that allows for the generation of ketones from carboxylate salts. The calcium carbonate gets recycled during the formation of a ketone and then gets added again to two carboxylic acids to help form a ketone again.

Carbohydrates – Biopolymer polysaccharide compounds contained in food crops such as starch or ones found in trees and plants such as cellulose.

Carbon dioxide assimilation (fixation) cycles – The incorporation of carbon dioxide that gets introduced into various genetically engineered metabolic pathways in photosynthetic types of microorganisms. The carbon dioxide then later becomes converted into other organic compounds of interest such as ethanol instead of normal compounds such as saccharides. The entry points of carbon dioxide assimilation are the **calvin cycle** and the **C4 fixation.** The carbon dioxide turns into other precursor compounds formed by the **citric acid cycle** until they become pyruvate.

Carboxylic Acids (Carboxylate salts) – Organic compounds that contain a carboxylate group and a hydrogen at the end of it. They are oftentimes also called **organic acids**. The alternate carboxylate compound (a negative anion resulting from the removal of hydrogen) is formed with the addition of another salt such as calcium carbonate or ammonium bicarbonate. The carboxylate salts are the chief initial products formed during the carboxylate counter current fermentation process. Some common carboxylic acids include *acetic acid, propionic acid, butyric acid, valeric acid, caproic acid.*

Carboxylate Counter-Current Fermentation – A biofuel production model developed by researchers at Texas A & M University that allows for the production of either mixed alcohols or hydrocarbons for direct use as a vehicle fuel or fuel additive. It is a biorefinery based production model that incorporates the main ideals of carboxylic acid production from the anaerobic digestion of renewable wastes by microbes. However, instead of producing methane these microorganisms produce carboxylic acids. The rest of the refinery processes either convert the carboxylic acids to alcohols through thermochemical methods or recycle the leftover biosolids and lignin material to make electrical energy or hydrogen for the alcohol conversion process.

Carrier gas – A type of effluent or recycled waste gas that is input back into the reactor part of the refinery in order to assist with the improved formation of products made within the reactor.

Catalyst – A material that increases the rate at which a chemical reaction takes place. In terms of biofuel production in this book a catalyst is implemented during thermochemical gasification, various production stages in carboxylate counter-current fermentation, liquefaction and the hydrotreatment production or refining method as well.

Catalytic synthesis stage – The process stage in thermochemical gasification whereby synthesis gas gets converted into alcohols or other hydrocarbons. In summary the synthesis gas components gather on metal catalytic surfaces. Afterwards they volatize off of the catalysts and then become condensed into a product solution that can be later processed.

Catalytic upgrading – A refining process applied to pyrolysis made bio-oil done in order to stabilize and reduce the oxygen content of the bio-oil so it can be later utilized as a vehicle fuel. It takes place in a separate reactor containing a metal catalyst executed at high temperatures.

Cellulose – A structural biopolymer of plant material consisting of glucose units glycosidically linked together. It is found as a crystalline material type matrix where glucose chains are bound together as fibrils that run across perpendicular to each other much like a scaffold structure. The majority of biomaterial contained in plants consists of cellulose. Cellulose is usually the main polysaccharide material found in lignocellulosics that can be fermented or converted into other chemicals.

Cellulosic Biofuels – A biofuel category classified by the Renewable Fuel Standards that qualifies a biofuel produced from a lignocellulosic source. The manufacturing methods usually include themochemical gasification and ethanol production. It is expected that 16 billion gallons of biofuel be made from lignocellulosics by 2022.

Cetane value – A measurable number determined by a standardized diesel engine test that utilizes a reference fuel. The engine testing conditions compare an optimum ignition point with the sample or test fuel. Cetane number evaluates the combustion efficiency of a given fuel or fuel compound.

Char – A solid byproduct made from pyrolysis or thermal gasification. It contains mainly elemental carbon. It can also be recycled into products or utilized in char combusters to generate heat or effluent gases during biofuel production.

Char combustor – A unit connected next to a fluidized bed reactor (i.e., gasifier) that heats char and alkali metals in order to transfer heat effectively to the fluidized bed reactor connected next to it.

Chemical building blocks – Raw biochemicals that when converted into other synthesized compounds assist towards forming the necessary ingredients contained in consumer and industrial products. They sometimes are also the main precursor chemicals that can help to make biobased polymer plastics such as polyamides, polyurethanes and polycarbonates. The government has classified 20 top biobased chemical candidates that have the possibility of being major chemical building blocks for major potential manufacturing based applications. Therefore, many of these chemicals could be excellent candidates for alternative byproducts made by an integrated biorefinery.

Chemical pretreatment method – A lignocellulosic pre-treatment method required before further saccharification and fermentation of either cellulose or hemicellulose material takes place. This method implements chemicals such as acids, bases and alcohols sometimes at high temperatures and pressures as well.

Chlorella sp. – A very common type of freshwater algae that has been involved in the production of biofuels. Since there are many different types of chlorella there are also a number of different production methods that produce biofuels depending upon how the algae are cultivated.

Chlorellins – A class of compounds naturally excreted from algae such as chlorella that have the potential to be fashioned into biofuel. These compounds are excreted as a defense mechanism of algae responding to bacteria that grow near it. Chlorellins contain several types of organic compounds with the majority of it consisting of fatty acids.

Circulating Fluidized Bed (CFB) Reactor – A fluidized bed technology that incorporates a mixture solid particles such as sand with a fluidizing agent such as steam or other gases. The combination of them allows for proper mixing and heating of biomass so that it degrades efficiently into synthesis gas or other organic volatiles.

Clostridia – A type of bacterial genera that usually grows in anaerobic conditions and is found in normal environments such as soils. It appears to be a very suitable type of microbe for ethanol production as it grows well and easily in many types of bioreactor environments that can utilize a variety of feedstock sources. This book describes the production of ethanol from *Clostridia* through the metabolism of synthesis gas and also lignocellulosic biomass, which usually produces acetic acid via fermentation. Some species of Clostridia mentioned for biofuel production purposes in the book include *Clostridium Ljungdahlii, Clostridium carboxydevorans, Clostridium lentocellum* and *Clostridium thermocellum* (all mentioned in chapter 5).

Cloud point – The temperature at which crystallization starts to take place with the biodiesel fuel mixture. This adversely affects the fuel flow and behavior of it in the engine. The cloud point of biodiesel for the different oilseed sources varies markedly between them sometimes. Therefore, even some types of biodiesel can start to crystallize above freezing temperatures.

Co-cultivation – A circumstance where two different microorganisms grow together in the same culture media. Oftentimes the presence of another microbe elicits the formation of other compounds from the other microbe.

Cold filter plugging point – The temperature at which the biodiesel starts to clumpen and therefore would block fuel lines and fuel pump.

Cold flow properties of biodiesel – When the temperature of biodiesel falls below a certain point the fuel changes chemically and physically so that it will not perform well in an engine. The fuel can thicken, clump, crystallize or can just stop flowing. The low temperature properties that describe these behaviors are pour point, cloud point and cold filter plugging point.

Combustion efficiency – An engines ability to transfer most of the energy from combustion into mechanical energy that is directed towards propulsion. In other words an engine that operates efficiently will have less of the energy being directed towards waste heat and more towards mechanical propulsion.

Combustion enthalpy – The consumption of fuel in an engine requires oxygen and can therefore be measured as an exothermic process (gives off energy).

Combustion stability – The consistency or steadiness of the combustion performance in an engine.

Commodity chemicals – Certain chemicals defined by the chemicals industry that are manufactured at very large volume amounts in order to satisfy market demands. These types of chemicals also tend to find numerous applications as ingredients that assist towards manufacturing a variety of consumer and industrial products. Most of the time these types of chemicals are usually petrochemically based but biochemicals are increasingly finding their way into further applications that can result towards their higher demand allowing them to be potentially classified as commodity chemicals.

Compression ratio – The ratio in volume measured between when the piston is at its uppermost position versus when its at its lowest position during a combustion cycle. Oftentimes experiments will perform engine tests based on a **variable compression ratio** meaning that a second piston is placed within the first one allowing for the compression ratio to vary during engine combustion cycles.

Condensation reactions – A type of reaction that joins two types of compounds into a larger one with the resultant byproduct formation of water. For example several organic acid compounds can combine together to form ketones. The ketones can also combine together to form larger compounds.

Conservation Reserve Program (CRP) – A conservatory program headed by the Farm Service Agency (FSA) of the US government that allows for the rehabilitation and improvement of pasturelands and farmlands located across the country. The projects are dispersed as 10 – 15 year contracts issued to landholders that attempt to grow crops such as grasses for land cover purposes in order to improve the water quality, prevent soil erosion and reduce the loss of wildlife habitat of such lands.

Consolidated Bioprocess (CBP) – A process that theoretically combines the process steps of saccharification and fermentation into one overall process ideally using one type of microorganism. This basically removes the need to add cultivated enzymes for the saccharification process step.

Corn starch ethanol production – Ethanol manufactured from the fermentation of corn starch in multimillion gallon per year refineries. The process separates the corn starch from the corn kernel in a dry grind corn mill. It simultaneously produces dry grain solubles as animal feed while the corn starch is saccharified and fermented using enzymes and natural S. Cerevisiae yeast cells.

Corn stock to use ratio – The amount of corn available when taking into account the amount produced compared to the amount stored or saved away. This metric helps to determine the supply and demand relationships as applied towards the commodity value of corn.

Corn stover – A highly plentiful crop residue leftover on the ground after corn harvesting. It mainly consists of the leaves and stalks of the corn plant. It is the largest crop residue available by far within the United States. Corn stover can also be utilized towards producing biofuels such as ethanol.

Corn sugar products – There are three major corn sugar based commodities sold on the market that are utilized commonly in food and beverage products. These commodity products are HFCS (High Fructose Corn Syrup), glucose syrup and dextrose.

Corrosive and hygroscopic nature of ethanol – Ethanol attracts and obtains water readily in its anhydrous state. In addition, due to its chemical properties, it is also corrosive in nature somewhat so that it makes it difficult to transport it in pipelines.

Crop residues – A renewable waste resource that consists of leftover crop material from harvests that can be collected each year. Types of crop residues include corn stover, wheat & rice straw, sorghum, sugarcane bagasse, citrus waste, etc. These residues are often utilized towards producing biofuel or other products such as chemicals.

Crystalline cellulose – The organized structured form of cellulose found in lignocellulosics or other biomaterial. The crystalline cellulose helps to form more complex scaffolding structures such as fibrils that help make up the macrostructure of cellulose in these types of biomaterials.

Cyanobacteria – Also known as **blue-green algae**. They are a photosynthetic microorganism that is prokaryotic in nature. Therefore, many consider them more a bacteria than an algae even though they rely upon photosynthesis to provide for their energy needs. They are probably the most common type of microorganism/algae that exists in natural environments that can be utilized towards producing biofuel. They are ideal for biofuel production since they are easy to cultivate, grow rapidly, form a variety of different end products and are amenable to manipulation through genetic engineering.

Cyclone separator – A piece of equipment installed next to a gasifier or reactor that effectively separates out large sized particulates that would normally get into the main product stream whether it is organic volatiles in pyrolysis or synthesis gas for thermochemical gasification. It usually separate out particulates greater than 5 microns in size.

Decantation – The act of pouring out part of a liquid contained in one vessel and transferring it into another one. This type of separation usually takes place with bio-oil that has had water added to it. It then separates into organic and aqueous fractions. The two fractions can be separated by simple decanting.

Dehydration – The chemical process of removing a hydroxyl (-OH) group from an organic compound. This usually results in the formation of an alkene (double bonded hydrocarbon). Alcohol compounds are usually converted into alkenes through dehydration.

Desert oilseeds – Certain oilseed crops that grow well in semi-arid or desert areas. They include types such as **Jatropha, castor beans, salicornia** and **jojoba**. These crops oftentimes grow on marginal lands and give high oil yields per acre, making them attractive to use as energy crops.

Diesel based biofuel additives – Synthesized organic compounds that have special properties that make them suitable additives for either biofuel based diesel or even petrodiesel. They can also be effectively utilized in biodiesel as well. These additives have excellent viscosity values and low temperature stability indicative of a compound that has a low melting point temperature. Some additive compounds include dibutyl ether, ethyl acetate, dimethyl carbonate and dimethoxymethane.

Direct ethanol production (from algae) – The process whereby algae produce and secrete ethanol as a byproduct of their own metabolism whether it be from autofermentation or carbon dioxide assimilation. A good number of algal species have been genetically modified to produce ethanol usually from carbon dioxide assimilation.

Direct to Ethanol® process – A proprietary production process developed by the company Algenol that in essence collects and separates the ethanol produced by genetically modified algae cultivated in a photobioreactor. The process collects an ethanol/water mixture that condenses on the top surfaces of the photobioreactor and then drains down into the troughs.

Distillate fraction – A set of hydrocarbons obtained from a thermochemical production method based on the boiling point range of compounds isolated through distillation. A **distillation column** gathers given sets of hydrocarbons into individualized areas that can be further processed as groups (fractions) to form refined chemicals or vehicle fuel.

Distillation – A unit operation that takes place in a refinery meant to separate a mixture of liquids based upon their vaporization points. The separation process involves the combination of evaporation and further condensation taking place within a **distillation column**. The column usually contains a set of miscible compounds in specific areas or collection plates where the liquids condense.

DOE Aquatic Species Program – A research and development program conducted by the DOE in cooperation with several research institutions located around the US that took place during the period from 1978 to 1996. The goals and objectives of the program were to ascertain the practicality of producing biofuel (mainly biodiesel) from the cultivation and processing of algae. It aimed to categorize the natural types of algae available in certain regions of the country that have the potential to make biofuel.

DOE 2009 Integrated Biorefinery Demonstration Project – A DOE based funding program meant to assist companies building biorefineries that meet the criteria of novel based integrated biorefinery technologies either on the pilot plant or demonstration level. The program came out in 2009 and awarded 19 recipient companies funded grants valued at several million to around 50 million dollars per refinery.

Effluent gas – These are exhaust type gases emitted from the catalytic synthesis stage of a thermochemical gasification refinery. Oftentimes these gases are recycled through transfer to other unit processes of the refinery. Effluent gases usually consist of methane and carbon dioxide.

Electrostatic precipitator – A piece of equipment utilized in a thermochemical gasification refinery in order to remove fine particulate material that is usually less than 5 microns in size. It accomplishes this by enforcing an electrostatic charge on particles passing through it so that they can be collected by charged plates located within the unit.

Elemental carbon – Inorganic carbon found in an amorphous state that is the main constituent in char resulting from thermochemical gasification, pyrolysis or liquefaction. The material usually just consists of carbon atoms bonded together.

Energy crops – Crops that usually are useful towards producing energy and products. Energy produced from them can apply towards electrical energy or vehicle fuel. They normally consist of lignocellulosic types such as certain trees or grasses. These crops are cultivated just like food crops located in specialized cultivation areas. However, they normally do not consist of food based crops. Various types of oilseeds also qualify as energy crops. A loose definition may even apply with algae as it can be cultivated and harvested having the same purpose as other energy crops.

Energy density of a fuel – The amount of energy stored in vehicle fuel measured by the amount contained in the hydrocarbon compounds per unit volume of the fuel.

Energy Indepedence and Security Act of 2007 (EISA)– Contains US renewable energy policies as applied towards electrical energy generation and biofuels production. It also contains past policies related towards the attempted establishment of ethanol production and infrastructure.

Entrained flow reactor (gasifier) – A reactor configuration for producing synthesis gas where the material reacts at very high temperatures and is input at the top of the reactor so that a slag material exits at the bottom of it. This type of reactor setup is mean for fine particle type material such as coal.

Escherichia Coli – A common type of bacteria used in genetic engineering that has the goal of making specific end products from modified metabolic processes. Any type of microorganism can be utilized in genetic engineering. However, Escherichia coli is the most common type that is worked with using such technology.

Esterification – An organic reaction that takes place between two organic compounds with the addition of heat that then forms an ester compound. Two types of esterification reactions are covered in this book. 1) The reaction between a carboxylic acid and an alcohol. 2) The reaction between a fatty acid and an alcohol.

Esters – An organic compound class that contains an overall structure of a carboxylate group (O-C=O) in between other normal hydrocarbon groups. Esters are made from the combination of a carboxylic acid (carboxylate salt) and an alcohol. They are one of the intermediate organic compounds made from carboxylate salts in the carboxylate counter current fermentation process used to make mixed alcohols. Many types of ester compounds have also been made and studied for future use as gasoline or diesel fuel additives.

Ethanol – A two carbon alcohol that is usually produced through fermentation utilizing saccharide based feedstocks. There are also other ways that ethanol can be made with thermochemical gasification being an example. Ethanol is currently our oxygenated based fuel additive put into gasoline. However, it is questionable whether it is our best option to use as a fuel additive due to concerns about how it is currently made (from corn starch) and adverse chemical properties that it contains. Ethanol could also become a valuable commodity chemical used in the future to help make other chemicals.

Ethers – An organic compound class that contains an oxygen atom in between other hydrocarbon groups. Ethers are usually synthesized from other organic compounds. It has also been the opinion of many that several types of ethers would make suitable fuel additives both for gasoline or diesel fuel. However, MTBE, another type of ether, was banned as our last gasoline fuel additive right before ethanol was instituted.

Ethyl acetate – A common ester compound formed during the esterification process that takes place with acetic acid and ethanol. It is the intermediate product formed during fermentative acetic acid conversion before hydrogenation happens in order to form a set of ethanol compounds.

Exopolysaccharides – Saccharides that are contained outside of the cell of both microbes and algae. They are excreted by the microorganisms to form an extracellular matrix or be found free flowing in the solution surrounding the microorganism. Expolysaccharides have been found to have special properties that make them candidate materials for ingredients in products such as surfactants and emulsions.

Fatty acid composition – Biomass sources that contain lipid biomaterial can be profiled as to the types and amounts of each fatty acid contained in the source material. This oftentimes applies for algae and oilseed sources. The most common fatty acids contained in lipid sources seem to be **linoleic, palmatic** and **oleic acids**.

Fatty alcohol – A compound similar to a linear alkane but has a hydroxyl group at the end of it. Fatty alcohols are good candidate biofuel compounds as they are in the same size range as methyl esters (from fatty acids). For example, they could be directly utilized as a diesel fuel compound without the need for having it refined.

Feedstock – A material supply of a renewable biomass or waste source that is utilized towards producing biofuel and/or some other product of value.

Fermentation – A biological metabolic based process that involves the actual conversion of sugar or saccharide compounds into ethanol or other chemicals using microbes such as yeast cells. Fermentation oftentimes produces more than one organic chemical product unless the participating microbe is genetically engineered to do otherwise. Fermentation utilizes saccharide based feedstocks such as starch, sucrose, cellulose, etc.

Fermentative acetic acid conversion – A hybrid ethanol production method that involves the initial production of acetic acid via the fermentation of lignocellulosic biomass. The acetic acid is then converted to the alcohol ethanol through thermochemical processing steps such as esterification and hydrogenation. This technology is very similar to the carboxylate counter-current fermentation process.

Fermenter – A bioreactor design specially dedicated towards cultivating a microorganism (i.e., bacteria or algae) or mixture of them in order to produce specific end products that could either be the microbial biomass itself or compounds excreted or contained within microorganisms. Fermenters usually have tight controls regarding temperature, pressure, aeration and mixing conditions along with the need for certain chemicals added in a periodic timely fashion.

Fermenter gas – The metabolic end products of fermentation using microbes in a fermenter. The gas consists of hydrogen and carbon dioxide. It is formed in synthesis gas fermenters as well as counter-current fermenters.

Fischer-Tropsch Synthesis (FTS) – A refinery method that essentially converts biomass, coal or natural gas into hydrocarbons that make up vehicle fuel or other chemical products. The method was invented by German scientists in the 1920's and works by having synthesis gas components react on metal catalytic surfaces to produce a large range of hydrocarbons. The process can either be split into a low or high temperature process. The hydrocarbon fractions normally require further upgrading or refining except for a straight run fraction which can be directly used as diesel fuel.

Flex Fuel Vehicle (FFV) – A spark ignition engine vehicle that has been modified with special linings and materials within the engine in order to operate on the fuel E85. Flex Fuel Vehicles can also operate on regular gasoline as well, thus the name Flex Fuel.

Flue gas – A waste or exhaust gas that exits through a large pipe from places such as manufacturing or electrical generation plants. The flue gas at electrical generation plants emanates from the combustion of coal and usually contains particulate matter, carbon dioxide and other contaminants.

Fluidized Bed Reactor (Circulating or Bubbling) – A fluidized bed technology that incorporates a mixture solid particles such as sand with a fluidizing agent such as steam or other gases. The combination of them allows for proper mixing and heating of biomass so that it degrades efficiently into synthesis gas or other organic volatiles. There are several configurations of this technology with two being circulating or bubbling fluidized bed reactors. See chapter section 3.4 for more information.

Forestry waste residues – Waste material that is available from the gathering and processing of trees used to make lumber products. It is potentially the most abundant renewable waste within the US that can be used to make energy, fuel or products. Forestry waste residues can be obtained from logging activities, forest clearing, removal of wood from forest fires, unmerchantisable wood and wood leftover from electrical energy generation. It can also include wood shavings, pieces or sawdust that originate from lumber, paper, saw mills as well as manufacturing outfits that make industrial wood products.

Fractionation – A pretreatment process that effectively separates cellulose and hemicellulose material from the lignin components of lignocellulosics. This method usually combines a thermochemical treatment along with solvent extraction to accomplish this.

Free fatty acid – An individual fatty acid that is not bound together as a more complicated organic compound such as a triacylglyceride. Free fatty acids are the precursor compounds produced during lipid metabolism that later on form highly organized structures contained in the cell wall of microorganisms. The presence of free fatty acids make biodiesel production from oilseeds or algae more complicated in that a separate processing step is usually necessary to execute in order to convert them into methyl esters. If this step is not carried out the compounds can form contaminants such as soap compounds that have to be separated out from the overall biodiesel solution.

Furfurals – Chemical byproducts during biofuel manufacture that can be later converted into a set of plastic related materials. They are often obtained from the conversion of hemicellulose sugars using heat and pressure. They are also compounds produced from the aqueous phase processing of saccharides that can be later converted into other chemicals.

Gasifier – The primary unit in a thermochemical gasification refinery that converts biomass or other material into synthesis gas plus other gases or impurities. It often combines steam plus biomass at high temperatures in order to facilitate a number of reactions that cause the production of primarily synthesis gas. Fluidized bed reactors are the gasifier units utilized in many refineries.

Genetically modified microorganism – In terms of biofuel related production there are many circumstances where changing a microbes regular metabolic processes favors the production of specific compounds that are suitable for biofuel use. This oftentimes involves the addition of other enzymes that direct metabolic intermediates towards the favored organic compound. This creates the ability to produce the candidate biofuel compound in larger quantities and yields.

Glycerol carbonate – A compound synthesized from glycerol via an etherification process. It is valuable in the fact that it is a component that can assist towards producing the biopolymer plastic compounds of polycarbonates, polyamides and polyurethanes.

Government monetary assistance programs (corn starch ethanol production) – These include government based funding, grants, loans and tax breaks and rebates that are (were) applied towards corn ethanol production for the producers.

Greenhouse gas emissions (GHGs) – The emission of gases that contribute to an atmospheric radiation effect. These gases oftentimes include methane or carbon dioxide. Most biorefineries produce carbon dioxide as a byproduct of processing whether it be from fermentation or a thermochemical based production method.

Heating value – The energy content of a given amount of fuel. It is usually measured in KJ per Kg. The value measures the amount of heat produced upon the complete combustion of a given amount of fuel.

Hemicellulose – A structural biopolymer of plant material consisting of straight chains and side branches of various 5 or 6 carbon saccharides like xylose, arabinose, galactose and mannose. Xylose is the most common saccharide found in hemicellulose known as xylan. This material consists of shorter rod like structures that attach to and help connect cellulose fibrils together. Hemicellulose similar to cellulose can be fermented by microbes or converted into chemicals through thermochemical methods.

Heterotrophic fermentation – A method of producing an accumulation of intracellular lipid compounds from the intermittent feeding of high concentrations of sugars or other types of organic carbon compounds. This fermentation is done within a short period of time (~ one to several days) and executed in the dark with a large amount of mixing involved in the process. Heterotrophic fermentation can take place with algae or microbes.

High Fructose Corn Syrup (HFCS) – A corn sugar product obtained from the conversion of dextrose into fructose executed with the use of microbes.

Homogeneous Charge Compression Ignition Engine (HCCI) – An engine design still under development that performs more efficiently due to effectively distributed homogeneous fuel mixture that allows for complete and even burning of the inputted fuel during combustion. In addition the engine design allows for tightly regulated controls such as appropriate air intake amounts, fuel injection timing and air-fuel mixture temperature. This engine design still has several drawbacks to overcome before being implemented on wide scale use.

Hybrid ethanol production method – An alternative ethanol production method that involves the combination of a thermochemical based processing with fermentation. Such technologies can include synthesis gas fermentation and fermentative acetic acid conversion.

Hybridization (variant genetic hybrid) – Applies to the interbreeding or genetic cross transfer of DNA between different genetic species. A set of genetic variants are produced when microbes or algae come into contact with each other as it is easy for them to share and transfer DNA between. Genetically modified microbes transfer their DNA to 'wild type' microbes allowing them to create 'hybrids' that have improved genetically based characteristics over the normal wild type microbes.

Hydrocarbon cut - A set of hydrocarbons produced from thermochemical gasification that usually fall within a certain size range that then gets processed by a specific upgrading method in order to produce usable vehicle fuel.

Hydrocracking – A refinery process whereby larger sized hydrocarbons are broken down into simpler types that can also rearrange themselves into usable fuel compounds such as isoparaffins. A hydrocracking refinery implements catalysts along with a supply of hydrogen. Hydrocracking normally is utilized with the larger sized hydrocarbons produced from thermochemical gasification when referring to biofuel production.

Hydrodesulfurization – A refining process that takes place with petrodiesel fuel meant to remove most of the sulfur content from it so that sulfur dioxide emissions from diesel fuel are no longer a concern. The catalytic refinery unit process that removes the sulfur is called a hydrotreater.

Hydrogenation of ketones or esters – The addition of hydrogen reacted with ketone or ester compounds further converts them into alcohols. Hydrogenation of an ester results in the formation of two alcohol compounds.

Hydrolysate – A solution containing broken down monomer based compounds that are the result of a hydrolysis process. These solutions are usually formed from a pretreatment process applied to either lignocellulosic or algae biomass utilized towards making biofuel and/or chemicals.

Hydrolysis – The act of breaking bonds between molecules such as saccharides with the resultant release of a water molecule. It often refers to the degradation of cellulose or hemicellulose material by breaking down the glycosidic oxygen bonds that bind glucose or other saccharide molecules together. This can be done through chemicals such as acids or with specialized enzymes.

Hydrotreatment – There are two ways to describe hydrotreatment
A) A post processing or refining method that converts oxygenated hydrocarbons contained in raw fuel cuts or bio-oil to normal hydrocarbon sized biofuel compounds accomplished through a number of reactions that take place in a confined environment. The process takes place in a reactor that contains a catalyst operated at high temperature and pressure along with the addition of hydrogen.
B) A fuel processing treatment that directly converts lipids, vegetable oils or fats contained in oilseeds, algae or animal waste to renewable diesel or jet fuel. The manufacturing conditions are very similar to the process described above. However, the fuel is produced in one processing step instead of going through an initial production method followed by the hydrotreatment process.

Hydrous ethanol – Ethanol that contains some amounts of water usually in the range of 5 % by total volume. This type of ethanol removes the need for a manufacturing related dehyradation step and therefore, should be cheaper to produce. It may also be more beneficial for vehicle use and its relevant transportation across the country.

Hydroxide catalyst – The chemical used to facilitate the esterification reaction between an alcohol and fatty acid that takes place during biodiesel production. The most common types of chemicals implemented as catalysts are sodium and potassium hydroxides.

Ignition delay – The time difference between when the a fuel is injected into the engine and the position just past the optimum point of combustion in a diesel engine. Ignition delay is a metric that determines the cetane number of diesel fuel.

Immobilized enzyme system – A production system that implements the use of enzymes supported on a material so that they do not move or flow free in solution. This production method has many advantages over enzymatic systems that are just placed into the reaction solution. Immobilized lipase production systems may allow biodiesel production to take place on an industrial scale.

Indirect ethanol production (from algae) – The process of making ethanol from algae where the algae mainly just supply the source of saccharides that yeast or other microbial cells use during fermentation to produce the ethanol. Its similar in concept to lignocellulosic fermentation where lignocellulosic biomass sources supply the saccharides but the yeast cells ferment them to make ethanol.

Industrial wastes – Waste materials or solutions that emanate from an industrial production process or previously used manufactured products. If the wastes are organically based they are potentially useful towards renewable biofuel and chemicals production. Some industrial wastes include biogenic – medical waste, leftover paints, inks, solvents, fluids (i.e., lubrication), and various types of sludges leftover from manufacturing.

Integrated Biorefinery – Modernized novel biorefinery technologies that companies have developed in order to competitively produce biofuel for potential and future market use. These types of biorefineries make efficient utilization of biomass towards producing biofuel, value added byproducts and power/heating needs. These types of refineries also significantly reduce carbon dioxide emissions. Integrated biorefineries are just starting to be built within the last decade and therefore, many of them are still in the pilot or demonstration phase of production.

Intermittent feeding cycles – An algal or microbial cultivation method meant to produce specific types of metabolic products by introducing a certain amount of nutrients that are taken up within a certain period of time. The cells then require further addition of these same nutrients later on after the supply of them has been depleted for a given amount of time.

Intracellular lipids – Lipids that are formed, stored and transported within the cell instead of being contained in the cellular membrane. These lipids are usually found as triacylglycerides pools contained in lipid inclusions or vesicles. These are specialized storage compartments found in the cytoplasm of the cell.

Isobutanol – A branched four carbon alcohol that has excellent chemical properties that make it an ideal candidate as a fuel additive in gasoline. It can be produced either through fermentation or thermochemical gasification. However, production via fermentation usually involves genetically modified microbes or algae.

Isobutyraldehyde – A specific metabolic end product made by genetically engineered cyanobacteria that can either intake carbon dioxide or bicarbonate as carbon sources. The isobutyraldehyde is similar in structure to isobutanol and therefore can be reduced through chemical or biological means in order to produce the isobutanol after it is made by the cyanobacteria.

Isomerization – A fuel refining process that can be applied to both petroleum and renewable based fuels where the initial hydrocarbons are cracked and then recombine to a number of different yet similar hydrocarbons. Therefore, many of the compounds produced have different structures but are of the same carbon length.

Isoparaffins – Hydrocarbons that are branched in structure and have no double bonds in them. They are a subset of alkanes. These compound types in part help to give gasoline its high octane rating. Refinery processes that include the reactions of alkylation, isomerization, dimerization, etc assist towards making a large percentage of these compounds in gasoline fuel.

Isopropanol – A branched three carbon alcohol that is produced from thermochemical gasification, one of the products from the ketonization process of carboxylate counter-current fermentation or fermentation sometimes involving genetically modified microorganisms. Similar to isobutanol, it also makes for a suitable fuel additive. Although higher molecular weight branched alcohols in general make for better fuel additives.

Jatropha – An oilseed crop that can grow in a variety of climates but is also well adapted to grow in desert areas as long as there is little to no frost conditions. It is a tree that grows large seeds with skins on them that have large oil content within them.

Jojoba – A plant that grows in arid regions of the world. It contains a seed that has large molecular weight waxy ester compounds in them. These esters can be broken down into alkyl esters and fatty alcohols either utilized for biodiesel or health care and cosmetic products.

Kerosenes – They are the principal compounds found in jet fuel and can be obtained from certain fractions of bio-oil or fischer-tropsch hydrocarbon cuts once they are refined in a certain manner.

Ketones – An organic compound class that contains an overall structure of a carbonyl group (C=O) in the middle or in between other normal hydrocarbon groups. They are one of the intermediate organic compounds made from carboxylate salts in the carboxylate counter current fermentation process used to make mixed alcohols.

Knocking of standard intensity – The octane number of a sample fuel or fuel compound is based on the quality of knocking intensity which is measured by a detonation meter. The octane number of the sample test fuel is determined when the vibrational pinging (knocking) of the sample test fuel matches the reference fuel.

Levulinic acid – One of the major biochemicals made from the aqueous phase processing of saccharides (glucose). Levulinic acid is very special in the fact that it can make other valuable chemicals or be further fashioned into biofuel compounds as well.

Life Cycle Analysis (Assessment) [LCA] – A methodology that assesses the environmental aspects or other effects associated with a production process. The biofuel production process itself usually calls for LCA studies as energy efficiency and reduction of carbon dioxide emissions are common aspects of production that have to be addressed and optimized.

Lignins – Biocompounds called monolignols are chemically connected to other plant polymers and dispersed throughout the plant matrix material that allow a plant its rigid structure and disallow predators such as insects from degrading plant material. Lignins are the biomaterial that cannot be fermented therefore other uses such as electrical power generation takes place after separating out the lignins during a pretreatment method.

Lignocellulosic biomaterial – This typically describes the feedstock source utilized to make biofuel and/or chemicals obtained from common plant sources such as trees or grasses. The term can also apply towards already processed materials that are usually wood product based. Lignocellulosics all share the same biopolymers that make up the majority of its core material. These substances are cellulose, hemicellulose and lignins.

Lignocellulosic ethanol fermentation – The manufacturing method described by using common lignocellulosic feedstocks that are later converted into ethanol via conventional fermentation. The process may involve conventional yeast or genetically modified microbes (i.e., other yeasts). In addition, other types of unit processes are employed that involve some type of complicated pretreatment method along with saccharification using enzymes.

Lime pile pre-treatment – A treatment method designed to separate the lignin from the cellulosic material. Large piles of lignocellulosic material are mixed with sludge/manure, lime and sometimes with bacterial innoculum as well. The piles remain as is for several weeks and then can be rotated amongst each other until the cellulosic material is ready to be digested in counter-current fermenters.

Lipases – Enzymes that break apart triacylglycerides and then facilitate an esterification reaction to take place between an alcohol and free fatty acid. Thus a biodiesel production process that uses lipases just requires the addition of an alcohol and no hydroxyide catalyst. Lipases as a term can describe the whole microbe that contains the enzymes or just the specific types of enzymes themselves.

Lipids – A class of organic compounds that are made up of fatty acids and derivatives of them. They preferably dissolve in hydrophobic solvents but usually not in aqueous based solutions. Natural biomaterial such as oils and fats are made up of lipids. This includes vegetable oils in plants and oils contained in algae or microbes.

Liquefaction – There are several ways to describe liquefaction

A. corn refinery & laboratory based algae – The process of heating and adding chemicals such as an acid to a solution of corn until the components of the corn kernel break apart and separate from one another. The same process also describes the method that separates the biocomponents of algae from one another using smaller scale laboratory equipment.

B. Themochemical processing method – A production process that has the end result of producing a bio-oil from high moisture content biomass. Liquefaction takes place usually in aqueous media at high temperature and pressure within a given type of specialized reactor.

Liquefied Petroleum Gases (LPG) – A set of small molecular weight hydrocarbon gases usually ranging from ethane to butane that are formed under high pressure in order to store them as a liquefied mixture. This term also applies to these particular gases produced and collected with Fischer-Tropsch synthesis process.

Low temperature hydrothermal processing treatment – A thermochemical refining method that slightly modifies algal or microbial biomass so that it can be separated into the components of saccharides, lipids and amino acids (polypeptides) later on. This type of processing is important in that it can make a type of biocrude or allow for the simultaneous production of biofuel from fermentation and lipid extraction. It also allows for the protein material to be separated for use as animal feed or fertilizer.

Lubricity of a fuel – Engine parts and components can wear down through abrasive contact that happens between the fuel and the parts themselves. Fuel therefore requires the addition of additives that improve the properties of the fuel so that they dramatically lessen the wear from physical contact. The fuel in essence is 'lubricated' so that less mechanical wear happens because of the vehicle fuel.

Lubricity wear scar (HFFR) – The size of an impression contained on a metal surface left by a metal ball immersed in a vehicle fuel that has a certain pressure exerted on it. The size is measured in microns and each fuel has a standard wear scare size that determines whether the fuel is 'lubricated' enough.

Macroalgae – Large aquatic plant like organisms that grow in large quantities at locations such as seas or oceans. They are a type of multicellular algae that can be readily seen by the naked eye. They consist of aquatic plant life such as seaweed or kelp. These algae are different in that they contain a large amount of unique types of saccharide material such as **laminarin, mannitol** and **alginate**. The term **brown algae** also usually refers to a type of macroalgae.

Major commodity food crops – The three main food crops of corn, soybeans and wheat cultivated in large amounts within the United States that have major economic and trade value. Together they make up greater than half of all available farmland utilized in the US.

Marginal land – Land that is unsuitable for common crop production use due to its poor soil conditions. Certain plants can still grow on marginal lands, sometimes even allowing for such lands to be rehabilitated for other uses. Certain types of oilseed or other energy crops can grow on marginal lands making them desirable as a source of biofuel since they do not compete for farmland or other valuable land space.

Melting point (freezing point) – The temperature at which a liquid compound crystallizes to form a solid compound. Many organic compounds that qualify as fuel additives have very low melting points meaning that they tend to help a fuel's cold temperature properties or assist the fuel to operate better in very cold conditions.

Metabolic branch point – A specific area in a metabolic pathway that determines what other types of chemical products can be made depending upon enzymatic activity and relative concentrations of precursor compounds. There is usually a choice in producing one of two possible compounds which then affect other branch points further down the metabolic pathway until the end metabolic product is made at the end of the pathway.

Methane – A one carbon gaseous hydrocarbon that makes up the majority component of natural gas or the main component of biogas. Methane in essence is the metabolic byproduct made by micoorganisms as they go through an anaerobic based digestion process associated with various forms of renewable waste.

Methanogen – A bacteria that emits methane as a metabolic byproduct while growing in anoxic conditions. Methanogens are common bacteria that appear in Anaerobic Digesters but are absent in carboxylate counter-current fermentation where the microbes produce carboxylic acids instead of methane as the main product resulting from the anaerobic digestion of biomass.

Methanogen inhibition – The process of preventing microbes called methanogens from growing in the anaerobic counter-current fermenters by adding chemicals or controlling the pH properly

Methanol – The simplest alcohol that consists of one carbon. This alcohol is made through thermochemical production methods rather than from fermentation. Methanol is a major chemical building block and can also produce further gasoline type compounds. Like ethanol it has adverse chemical properties that make it difficult to use as a fuel additive. However, through synthesis it can make other better oxygenated fuel additives.

Methyl esters – The chemical components that make up biodiesel when a lipid or oil source is processed with methanol. Methyl esters are formed upon the esterification reaction that takes place between a fatty acid and methanol with the aide of a catalyst.

Methyl Tert Butyl Ether (MTBE) – An ether compound that was utilized as our last oxygenated based fuel additive. It was banned in 2004 due to environmental concerns. It is made from the combination of methanol and isobutylene through reactive distillation.

Methyltetrahydrofuran (MTHF) – This is a cyclic oxygenated organic compound that can be derived from levulinic acid. It is valuable in the fact that it can currently be utilized as another oxygenated gasoline fuel additive.

Microbe – Various classes of microorganisms that are usually not photosynthetic in nature. This usually describes the microorganisms of bacteria, yeast and fungi but not algae (or cyanobacteria).

MixAlco® production process – An advanced bio-refining technology that converts low cost, readily available, non-food, non-sterile biomass into valuable chemicals such as acetic acid, ketones and alcohols that can be processed into renewable gasoline fuels. [** Definition directly taken from Terrabon's website] – Basically it is a patented version of the carboxylate counter-current fermentation technology that has been applied by the company Terrabon to make vehicle fuel and chemicals.

Mixed oxygenates – Oxygen based hydrocarbons produced from the Fischer-Tropsch fuel manufacture process. These compounds could be alcohols, ethers or other types.

Municipal waste – Solid waste derived from the consumption of consumer and industrial products. Most municipal waste is material that we consider trash and usually gets placed in a landfill. Some of the material is recyclable and can be turned into other products or energy. Most definitions mainly just include refuse as municipal waste. However, in loose terms it can be considered most types of recyclable waste emanated from urban types of facilities that can include restaurant food and grease waste, sewage sludge from wastewater treatment and some industrial types of waste.

Naphtha – A set of smaller sized hydrocarbons usually ranging from 5 to 12 carbons in length that arise from thermochemical gasification or pyrolysis. The naphtha can be separated out from the other types of hydrocarbon fractions and then get further refined to form a vehicle fuel such as gasoline.

Net Energy Value (NEV) – This is defined as the energy that remains as net fuel use after taking the difference from the amount of energy that was required to manufacture the biofuel from the actual amount of energy it contains sometimes also known as its heat value.

Neutralization – The process of removing free fatty acids from a batch made biodiesel solution or removing the free fatty acids in vegetable oil being processed into cooking oil. Both processes convert the free fatty acids into other compounds that can be removed from solution later on. However, both processes are executed a bit different from one another.

Nitrogen heterocyclic compounds – Cyclic aromatic compounds containing a nitrogen atom within them. There are a large number of compound classes that make up these type of compounds. They are usually some of the chemical byproducts found in bio-oil obtained from pyrolysis and liquefaction.

NREL indirect gasification mixed alcohols & ethanol refinery model – This was a DOE based pilot plant project that aimed at making ethanol and mixed alcohols from thermochemical gasification. It had unique design feature such as a two distillation column setup and indirect gasification from a char combustor. More details concerning the design of this biorefinery can be found on the internet as a DOE technical report entitled '*Themochemical ethanol via indirect gasification and mixed alcohol synthesis of lignocellulosic biomass*'

Nutrient depletion or limitation – Algal or microbial based cultivation methods that tend to increase the amount of intracellular lipids contained in the cultivated cell. This method completely removes the nutrients of interest (depletion) or drastically reduces the concentration of them (limitation) when executed as a second or additional cultivation stage. The removal of certain nutrients such as nitrogen forces a change in metabolism that produces more lipid compounds rather than others such as proteins.

Nutrient media – An aqueous solution that contains specific chemicals necessary to successfully cultivate a certain type of microorganism (i.e., algae, fungi or bacteria). The nutrient solution usually contains a carbon source, macro and micronutrients prepared at a certain pH.

Octane number – A measurable engine combustion performance value that is experimentally derived through engine tests. These tests compare a compounds standard knock of intensity with an already established reference fuel consisting of n-heptane and iso-octane. Octane number usually applies towards individual fuel compounds while **octane value** is the octane number of a given fuel mixture.

Oilseed cake – The material or meal leftover after the oil extraction process has taken place with an oilseed. The seedcake can be edible for consumption or contain toxins in it that make it unedible. The seedcake also has some amounts of residual oil contained in it.

Olefins – Organic hydrocarbon compounds that contain double bonds in them. They have the same meaning as alkenes. However, olefins are a term more applied towards both chemicals utilized in products or in vehicle fuel. They give a higher octane rating to gasoline than just normal paraffins.

Organics and food waste – This is a type of municipal waste source that emanates from any facility that prepares and serves food including places such as restaurants, grocery stores, institutions (i.e., universities, hospitals) and including but not limited to residential homes. Organic waste can include liquefied solutions of previous food matter. This type of waste is collectable and potentially can be converted into fuels and/or energy.

Oxidative stability – This term has various meanings dependent upon the specific chemical context. When applied to biodiesel it is the measurement as the amount of time it takes for the biodiesel compounds to degrade or chemically break down. Therefore, the oxidative stability of biodiesel helps to determine its shelf life or how long it can be used.

Oxygenated organic compounds (also oxygenated hydrocarbons) – Hydrocarbon based compounds that also contain functional groups consisting of oxygen. They are oftentimes grouped into compound classes such as alcohols, ethers, esters, carbonates and ketones. These compounds are normally synthesized from other types. In some circumstances these compounds are often utilized as fuel additives put into gasoline or diesel petro based fuel.

Paraffins – Hydrocarbons that are linear or branched in structure and usually have no double bonds in them, basically the same meaning as alkanes. However, they are oftentimes utilized more when referring to hydrocarbons contained in vehicle fuel or chemicals used in common industrial or consumer products.

Partial oxidation – One of the essential chemical reactions that take place in a gasifier undergoing thermochemical gasification. Essentially the hydrocarbon compounds contained in biomass react with oxygen executing partial combustion in order to form both the synthesis gas components of carbon monoxide and hydrogen.

Particulates – Tiny pieces of inorganic material that volatize into the product air stream formed during the heating process of thermal gasification. These particulates are considered contaminants and must be removed with specialized equipment so that they don't become part of the final product synthesis gas sent to the catalytic production stage.

Pentose and hexose saccharides – The majority of saccharides contained in biomass are usually made up of 5 (pentose) or 6 (hexose) carbon sugars. Hemicellulose and exopolysaccharides contain a large variety of saccharides such as **arabinose, galactose, fucose, xylose, mannose** and **rhamnose**.

Perennial crops – Crops that are known to grow or last for more than just a few years. The period of their harvesting can vary depending on the type of crop. For example some crops can be harvested once or twice a year but keep growing back and do not require replanting in order to grow again. These types of crops normally apply to trees or grasses.

Phenolics – When lignocellulosic material is broken down with thermochemical methods such as pyrolysis, they tend to form a variety of aromatic compounds that contain a hydroxyl group in them. The majority of these compounds come from the lignin components of lignocellulosic material.

Phospholipids – A type of lipid compound that makes up the main component of the lipid bilayer that is part of a cellular membrane. Phospholipids are made up of a phosphate group, a diglyceride part and another organic molecule such as choline. Phospholipids are usually not the main lipid compounds of interest when trying to make biodiesel from lipid material.

Photanol approach – A theoretical genetic engineering research platform meant to design algae that have the ability to make several types of alcohol products at one time from photosynthesis. In addition the algae should also be able to emit large quantities of hydrogen as well. The theory bases its work on the production and efficient use of **Glyceraldehyde-3-phosphate**.

Photobioreactor – A type of reactor similar to a fermenter that cultivates photosynthetic microorganisms but is designed to operate on the ability of the microorganisms to absorb light efficiently. Therefore, these reactors utilize clear, translucent materials and maximize light intensity and penetration into the culture solution. Similar to fermenters these reactors also require nutrient solutions, mixing apparatus and/or aeration systems that supply gases such as carbon dioxide.

Pilot plant – A small production facility that is built in order to help design the further construction of a biorefinery that will be able to produce biofuel at a large volume capacity such as multi-millions of gallons per year. The pilot plant usually operates at the capacity of many thousands of gallons per year. If a pilot plant is successful then the next step is to construct a **demonstration plant** that happens to be an order of magnitude or two lower in volume than the expected full scale production plant.

Plasma Enhanced Melter (PEM®) – A proprietary system developed by the company InEnTec that allows a facility to process municipal or other industrial wastes into synthesis gas. This is accomplished through a plasma melter system that 'melts down' the material with a high temperature charged plasma gas. It also forms vitreous solutions that can be recovered as metals or glass.

Polyhydroxyalkoanates (PHAs) – Organic compounds formed within microbes or algae that are suitable towards producing a bioplastic. Along with PLA it is one of the main bioplastics that are now commercially available. **Polyhydroxybutyrate (PHB)** is the main type of PHA found within microorganisms.

Polylactic Acid (PLA) – A bioplastic made from the polymerization of lactic acid based polylactide units. The lactic acid is normally produced from the fermentation of corn starch by microbes such as lactic acid bacteria.

Polysaccharides – Large polymer compounds found within the cells of animals, plants and microorganisms that help make up vital structural components of the cell material. Polysaccharides consist of individual or monomer units of saccharides linked together. They can consist of just one type of saccharide such as glucose that is found in cellulose or starch or a set of different saccharide units hooked together as is the case with hemicellulose.

Post reacted slurry – The overall mixture of char and bio-oil emanating from a pyrolysis or liquefaction based reactor. The slurry is processed after it leaves the reactor meaning that the char and bio-oil components get separated from one another in other unit process stages in the refinery.

Post sorted municipal waste – The municipal waste material leftover after it has gone through either recycling or composting operations. Recycling of waste oftentimes requires that the waste be collected, sorted and separated. The material that isn't recycled or composted can be considered post sorted municipal waste.

Pressure Swing Adsorption (PSA) – A method used to remove a gas from a mixture of others based on pressure and a specific adsorbent material that the gas can collect onto. The PSA unit is an important piece of equipment that has been utilized to remove carbon dioxide from other gases in a biorefinery.

Primary & secondary milling – Biomass feedstocks available for biofuel production must first be processed in order to make it suitable for introduction into gasifiers or reactors. Principally the biomass must be shredded, cut, grinded and sieved in order to reduce the material to the appropriate size range. Therefore, the biomass may require several processing steps usually requiring it to go to several milling facilities or it may just need to be processed in a single milling facility before it goes to the refinery.

Producer gas – A fuel gas that usually consists of synthesis gas and other gas components. It normally is combusted later on in order to produce energy for a refinery. Producer gas applies to a gas product made during pyrolysis or a portion of it made from thermal gasification that can be directed towards energy production.

Protein hydrolysis – The breakdown of proteins into smaller peptides or amino acids. **Protein metabolism (amino acid metabolism)** is not a process that usually results in compounds that are later used for biofuel production except in certain circumstances such as metabolically engineered biochemicals or the production of certain organic acids.

Pyrolysis – A method used to produce a bio-oil that becomes later upgraded to a renewable biofuel. Pyrolysis takes place in a reactor where biomass is heated for a certain duration until it reaches a certain temperature (around 400 – 500 °C). Thereafter the organic vapors produced from the reaction vessel go through a quick exposure period called residence time. Shortly after this period the vapors are quickly quenched (cooled down) in order to form a bio-oil. Pyrolysis can be classified as slow, fast or flash dependent upon residence time. Pyrolysis also produces other products like gases such as producer gas and carbon dioxide along with a solid waste product called char.

Pyruvate – A three carbon ketone type intermediate metabolic compound that is common to most types of metabolic processes including fatty acid, organic acid and alcohol synthesis. Pyruvate can be considered the major metabolic branching point when determining whether one or several alcohols can be made at one time from microbes or algae.

Pyruvate decarboxylase – One of the enzymes usually contained in yeast cells that allows for the conversion of pyruvate into ethanol through the intermediate production of acetaldehyde and carbon dioxide. This enzyme has been genetically transferred to algae cells so that they can also efficiently convert pyruvate into ethanol.

Rapid Thermal Processing® (RTP) – A type of fast or flash pyrolysis that has been developed by the company Ensyn. It describes the overall pyrolysis and upgrading process that can make a variety of fuels from a number of different feedstocks in an efficient manner, thus giving a large yield of fuel per ton of material input into the refinery. The company UOP uses this type of reactor technology to produce fuel at their pilot plant facility in Hawaii.

Reactive Distillation (RD) – A unit process contained in a refinery where the chemical synthesis and distillation steps are all executed in the same reactor. The synthesis of the ethers – MTBE or ETBE can be produced from Reactive Distillation.

Reference Fuel (octane testing) – A fuel for testing purposes that has a predetermined octane number resulting from the mixture of the two fuel compounds of n-heptane and iso-octane in any proportion. The reference fuel mixture that matches the best 'knocking of standard intensity' for a given compound will determine its octane number.

Refinery Capital Costs – One time, fixed costs that are incurred upon the construction of a refinery. Therefore, these costs usually include equipment, initial labor and other types. Oftentimes the overall capital cost is figured into the total price of producing biofuel per gallon by averaging the number of years it takes to pay off these costs along with the amount of gallons made by the refinery per year.

Refinery Operating Costs – The costs of various essential resources required to keep a refinery running on a daily basis. This oftentimes includes the cost of daily labor, utilities, chemicals and other expenses. Operating costs can be fixed or variable, meaning that they may change or remain the same during refinery operation.

Reformulated gasoline – A gasoline fuel blend that oftentimes contains a certain amount of oxygenated fuel meant towards improving both the clearness of combustion and emission standards of the fuel compared to just a conventional gasoline mixture.

Renewable biofuel – The more applicable definition applies towards the production of vehicle fuel from any renewable biomass or municipal waste source. However, the US Government's definition is mainly applied towards the production of ethanol from corn starch. This is one of the four biofuel categories classified by the Renewable Fuel Standards. The standard allows for 15 billion gallons of ethanol to be made per year up until 2022.

Renewable diesel – An umbrella term that includes any biofuel that can be operated in a diesel engine made from renewable sources. It is all inclusive describing the many ways that the diesel fuel can be produced from a variety of biomass sources. Therefore, this term could include the fuel of biodiesel although oftentimes people refer to the two as distinct types of diesel due to the difference in chemical compound makeup between the fuels. Similar to petrodiesel, renewable diesel usually contains a large variety of hydrocarbons that range in size usually from 10 - 20 carbons in length. Renewable diesel may be the most prevalent and convenient type of biofuel to manufacture. Renewable diesel also denotes that a further refining step such as hydrotreatment is required after its initial production.

Renewable Fuels Association (RFA) - An organization that highly supports the implementation of ethanol as an alternative biofuel and has assisted in the process of allowing ethanol to be used as a biofuel in vehicles on a large volume scale.

Renewable Fuel Standards (RFS2) - Mandated regulations set forth by the US government that stipulate the amounts of certain biofuels produced within the country. The regulations also qualify refinery operations by stipulating the amount of carbon dioxide that can be emitted by a refinery choosing to produce a certain biofuel. The standards classify a prospective biofuel into one of five categories. The mandates are oftentimes enforced by the EPA.

Renewable Identification Number (RIN) - A 38 digit code with set values used for production accounting purposes for biofuel manufacturers that is examined by the EPA on a yearly basis. Any company that produces over 10,000 gallons of biofuel per year must accommodate for RINs attached to every gallon of biofuel they produce.

Renewable Wastes - Materials that are waste products from consumer, industrial or farming applications but are regularly available for recycling into energy, products or fuel. There are several classes of renewable wastes that include municipal waste, crop residues, forestry waste, sludge from wastewater, waste vegetable oil and/or grease, animal wastes (i.e., manure or fats), etc. These waste sources do not include cultivation of algae, microbes or energy crops. The US government has ascertained that a large amount of energy, fuel and products can be made from renewable wastes (see US billion ton study for more information). Therefore, the proper utilization of renewable wastes is vital towards providing for our future fuel and energy needs.

Reverse water shift reaction - A chemical reaction that happens within a thermal gasifier that helps to form one of the synthesis gas components, that being carbon monoxide. Basically the synthesis gas component of hydrogen reacts with carbon dioxide in order to form carbon monoxide and water.

Saccharification – The breakdown of saccharides into simpler singular sugar units such as glucose through the process of hydrolysis that can take place through the use of chemicals and heat or enzymes.

Salicornia – A desert oilseed crop that especially grows well in sandy or marshy soils. It can also be cultivated with seawater. It produces a very healthy seedcake after the vegetable oil has been extracted from it.

Secondary (tar) reformer – Equipment used in the thermal gasification production process that recycles gases or breaks down compounds such as tars into synthesis gas components. The reformer can typically recycle effluent gas emanating from the catalytic synthesis stage.

Sewage sludge – Biosolids that accumulate and are separated out from wastewater at a sewage treatment plant. Much of the biosolids are a combination of leftover waste material along with microbial biomass that feed off of it. In the United States, the sludge usually contains toxic chemicals, heavy metals and possibly pathogens from industrial waste therefore it is normally incinerated and not applied to farmland as fertilizer. Even though it has toxic characteristics sewage sludge can be potentially converted into biofuel from a number of methods.

Solvent (hydrocarbon) extraction – The process of transferring a chemical component contained in a certain solution to another solution that is made up of an immiscible solvent. In essence the component dissolves into an added solvent that can be separated from the original solution due to the inability of the added solvent to be contained within it (i.e., immiscibility).

Specific fuel consumption – There are different types of specific fuel consumption. For most engines it is the amount of fuel required to move a vehicle a certain amount of thrust. In some ways people also think of this term measuring the specific amount of fuel required to have a vehicle travel a certain distance.

Steam reforming – One of the two essential chemical reactions utilizing the biomass itself that takes place in a gasifier during thermochemical gasification. Basically water vapor (steam) reacts with hydrocarbons or elemental carbon in order to form both the synthesis gas components of hydrogen and carbon monoxide.

Steam to biomass ratio – The amount of steam per biomass unit contained in a gasifier that helps to determine the consistency of synthesis gas components and impurities.

Straight run diesel fraction – A hydrocarbon component of a fischer-tropsch cut obtained from the high temperature process. This fraction can be directly utilized as diesel fuel once separated from the other components. It usually consists of hydrocarbons ranging from 11 to 22 carbons in length.

Synthesis gas – A gaseous mixture made up of two components with those being hydrogen and carbon monoxide. The mixture or ratio of these gases can vary depending on the associated manufacturing application. The mixture of these gases allows for a large variety of hydrocarbon compounds that are synthesized on metal catalysts.

Synthesis gas fermentation – A hybrid ethanol production method that involves the generation of synthesis gas from biomass and then the subsequent fermentation of the synthesis gas components by microbes that in turns produces end products such as ethanol.

Synthesis gas ratio – Certain types of thermochemical refinery operations require that the synthesis gas have specific temperature, pressure and chemical consistency of the inputted synthesis gas. This chemical consistency is known as synthesis gas ratio and is usually measured as the molar ratio of carbon monoxide to hydrogen. Depending on the application the synthesis gas ratio can usually vary from 0.5 to 2.0. Low and high temperature Fischer-Tropsch synthesis require differing specific values of synthesis gas ratio as does mixed alcohol or ethanol production as well.

Tallow – Animal fat obtained from cattle after it has been processed at meat rendering facilities. It can be processed into biodiesel or more oftentimes becomes an important ingredient in shortening.

Tars – High molecular weight aromatic hydrocarbons that are volatile. They are normally formed in a thermochemical based refinery during processing of the biomass in a gasifier. These compounds oftentimes consist of thermally degraded lignin compounds from lignocellulosic biomass.

Tall oil – A viscous liquid produced from kraft mill wood pulp manufacturing. Tall oil mainly consists of rosins and fatty acids. It can be further processed to produce biofuel and/or chemical products. The byproduct of tall oil is obtained from separating it out from the black liquor.

Tetrahydrofuran – A non toxic green solvent that has been utilized in biodiesel manufacture since it is similar to methanol and can be processed with it.

Themochemical gasification – A thermochemical based production method that converts biomass or other materials into hydrocarbons that can be later upgraded to vehicle fuel. However, this method can also specifically produce alcohols such as ethanol, which can be directly utilized as vehicle fuel. The method operates by mixing an already processed small particle size material in a gasifier with steam, oxygen or other fluidizing agent so as to cause the formation of synthesis gas. The other unit processes of the refinery must then clean and purify the synthesis gas product stream, recondition the synthesis gas to the proper temperature and pressure and then convert the synthesis gas to the appropriate hydrocarbon products through a catalytic synthesis stage. Fischer-Tropsch synthesis is a type of thermochemical gasification based technology but is normally the main type implemented to produce diesel fuel.

Triacylglyceride – A large ester compound found within animal, plant and microbial cells that happens to be the main constituent of fat or lipid cell structural compounds. It contains three fatty acids hooked together by a glycerol portion. Triacylglycerides are the main compounds of interest contained in oilseeds, algae or animal fat that become converted into methyl esters upon reaction with an alcohol and catalyst.

Two step catalytic biodiesel production method – More modern continuous large scale biodiesel production facilities execute a two step process that involves one type of catalyst converting the free fatty acid portion into methyl esters. Once these compounds have been removed then a second step converts the normal triacylglycerides into methyl esters using another catalyst.

Ultra low sulfur diesel (ULSD) – Diesel that has been upgraded through a hydrodesulfurization process meant to remove the sulfur in the fuel. ULSD has less than 15 ppm of sulfur contained in it. It has been the new petrodiesel standard since 2004.

Ultrasonic biodiesel production method – A very modern continuous type of biodiesel production method that utilizes ultrasound (i.e., sonication) to cause an esterification reaction to take place with the oil source of interest. The esterification reaction takes place very rapidly when exposed to the ultrasound and also expends much less energy for the conversion when compared to other biodiesel production methods.

Unburned hydrocarbons – Carbon compounds that are not combusted in an engine and therefore leave it through the exhaust system into the environment or go into the lubrication system.

Unmerchantisable wood – Lumber material that cannot be utilized for industrial wood products. It is usually made up of poor grade types of trees or bad sections of the trees. This biomaterial is one of the more readily available types of forestry residues usable for biofuel production.

Vapor Compression Steam Stripping (VCSS) – A set of refining processes applied towards separating and purifying ethanol at a refinery or plant. It usually combines a permeable membrane separation system with a steam stripping column. This method is effective at collecting ethanol from solutions that contain a low concentration of ethanol in it.

Vapor condensate fraction – A liquid fraction obtained from the condensation of small hydrocarbon gases found in the gaseous product stream made from pyrolysis. This condensate is separated from other gas types such as producer gas, carbon dioxide and effluent gas.

Vapor lock – The interruption of flow through a fuel line due to the vaporization of the fuel. The presence of vaporized fuel air bubbles tends to block the fuel lines.

Vegetable oil processing – The processing of vegetable oil takes place in refineries in order to clean, purify, deodorize and decolorize the oil so that it meets cooking oil applications. This involves the processes of degumming, bleaching, neutralization and deodorization.

Viscosity value – The measurable value of a liquid used to ascertain its ability to resist a change in flow conditions. This value is important for biofuel compounds as its flow characteristics determine the ease in which the compound can be transferred through the vehicle. It also helps to indirectly determine the volatile nature of compounds as they go through combustion. Compounds that do not easily volatize before combustion tend to form unburned hydrocarbons afterwards. Therefore, compounds with higher viscosities tend to not volatize as well and can favor the formation of unburned hydrocarbons.

Volatile organic compounds – The organic compounds produced within a pyrolysis reactor during its heating period and subsequent release from it. These compounds later on get converted into bio-oil after rapid quenching. The majority of volatile organic compounds consist of oxygenated hydrocarbons.

Waste to Energy (WTE) facility – A facility that basically produces electrical energy from the combustion of municipal waste. Most of the time the municipal waste is composed of recycled material such as cardboard, paper, plastic, etc.

Waste glycerine – The liquid byproduct leftover after the esterification reaction has made alkyl esters. This fraction can be drained out from the biodiesel portion. The waste glycerine fraction not only contains glycerol but other compounds such as leftover alcohol, soaps, acid and other contaminants.

Waste heat – The large amount of leftover heat usually generated from thermochemical gasification that can be utilized to heat water in a boiler to produce steam for electrical energy generation.

Water shift reaction – A chemical reaction that happens within a thermal gasifier that helps to form one of the synthesis gas components, that being hydrogen. Basically the synthesis gas component of carbon monoxide reacts with water in order to form hydrogen and carbon dioxide.

Waxy hydrocarbons – A set of hydrocarbons produced from the low temperature Fischer-Tropsch synthesis process. These hydrocarbons are normally greater than 20 carbons in size and can be utilized to make other chemicals or get converted into diesel fuel through an upgrading process such as hydrocracking.

Wet scrubbing – A process where scrubber equipment removes particulates, ammonia and other impurities from synthesis gas through a condensation mechanism using cooling water.

Yellow grease – A partially refined grease obtained from either waste animal fats or waste cooking oil. Waste cooking oil unprocessed usually contains a lot of contaminants it. Some processing removes many of these and makes a more usable product that can be utilized for biofuel or the manufacture of other products.

Yield per acre – The amount of biofuel or crop biomass produced per acre of land when made from an energy crop or other source such as algae. The amount of biomass is expressed as tons per acre while the amount of biofuel is gallons per acre. Biofuel yield per acre can be calculated by multiplying the biomass yield per acre by the biofuel yield per ton.

Yield per ton – The amount of biofuel produced ton of biomass it was produced from. This value is expressed in terms of gallons of biofuel per ton of biomass.

Zeolite catalyst – A microporous aluminosilicate material commonly implemented as a catalyst contained in biofuel refining reactors.

www.ingramcontent.com/pod-product-compliance
Lightning Source LLC
Chambersburg PA
CBHW082301200526
45168CB00017B/2298